The Best of Crocus Coulee

Betty Kilgour

Detselig Enterprises Limited
Calgary, Alberta

© 1986 **Betty Kilgour**
Crocus Coulee, Alberta

Canadian Cataloguing in Publication Data

Kilgour, Betty.
　The best of Crocus Coulee

　ISBN 0-920490-54-9

　I. Title.
　PS8571.I43A16　　1986　　　081　　　C85-091526-0
　PR9199.3.K54A16　　1986

First printing 1986
Second printing 1987
Third printing 1990

Detselig Enterprises Limited
P.O. Box G 399
Calgary, Alberta T3A 2G3

All rights reserved. No part of this book may be reproduced in any form or by any means without permission in writing from the publisher.

SAN 115-0324

Printed in Canada　　　　　　　　　　　　ISBN 0-920490-54-9

Introduction

Betty Kilgour spent her early childhood west of Sundre, Alberta, which was then a wilderness frontier. The family soon moved to small-town Alberta, out of the bush to the prairie — where there were crocuses! She grew up during World War II and then married Bill Kilgour, farmer and rancher. In the next twenty years, Betty helped raise grain and cattle — and five children. She drove the farm truck, hauled grain, kept house, hunted antelope, made mountains of lunches, fought mud, flies, hail and blizzards, nursed sick calves, delivered difficult-birth calves, herded cattle (on horseback in the Forest Reserve, and with a broom around the farmyard — and garden!) And did all the wonderful things a mother does for, to and with her children. And much more. And she wrote it all down.

If you are someone who sees sharply what others see only dimly, who experiences the full impact of circumstances — and then puts it into words so that the rest of us can also see and feel, laugh and cry, can revel in victories and share the despondency of defeat — and love it all — then you're a great poet. And if you can write in crisp, staccato sentences, with a sudden humorous word-twisting, depicting yourself as protagonist against all manner of forces, some evil, some just baffling, and you can come out victorious — more or less — then you've written a great book.

Such a book is *The Best of Crocus Coulee,* which was written over many years, but can be read in one sitting. I recommend doing just that. But afterwards keep it handy to read the little stories, one or two at a time, again and again.

Frank Jacobs

Acknowledgments

Heartfelt thanks to Peggy, Judy, Pat, Ian and Beth for providing so much raw material over the years; to Ted Bower for having faith in me for my ten years with the *Red Deer Advocate*; to the many provincial papers who carried my column, and to the editors of *The Western Producer,* the *Country Guide,* the *Cattlemen,* the *Canadian Hereford Digest,* Maine Anjou Magazine, and *Alberta Farm and Ranch* for using my material. Thanks also to Ben Crane for providing the illustrations and the cover artwork; to Ted Giles, my publisher, for believing it would work; and to Frank Jacobs for providing the Introduction.

Illustrations and cover artwork by Ben Crane
                                    Three Hills, Alberta

This book is dedicated to Bill, my partner at *Crocus Coulee*, with love.

Contents

Dr. Kilgour - Vet At Large 1
Sad Irons Were Sad! 3
Boy It's Cold! 5
Old Wash Days 7
Winter Styles - How They've Changed 9
Winter Fun 11
Christmas Long Ago 14
Diane Wasn't The Only One! 16
So You're To Marry A Farmer! 18
Now For Some Directions! 20
Please Take A Bone 22
Hair - The Crowning Glory 24
Riding Herd 26
Soda Isn't New 28
My Old English Bike 30
Crocus Time 32
City Cousins and Us 34
In The Ring 36
My Old Skates 38
The Old Back Door 40
The Eternal Triangle 42
Lunches - Lunches 44
Rural and Urban 46
Hey - The Roses Match! 49
On Getting Up 51
Income Tax Blues 53
How Men Change 55
Machinery and Me 57
Trucking - If you Dare 59
Harvest Time 61
Say "Cheese"! 62
Move Over, Dolly! 64
Don't Marry In The Fall 66
Sitting Duck 68
Heat The Irons, Boys! 70
Common As Dirt 72
Homemakers Don't Work! 74

The Queen's Mail 76
Mother's Gone A-Hunting 78
Housekeeping The Easy Way 82
Skunk Trouble 84
The Science Fair 86
Chinese Sign of War 88
New Wheels 90

Dr. Kilgour — Vet At Large

Well folks, just call me Dr. Kilgour, vet at large! I'm sure I've earned this title over the past years. Each and every spring I've assisted heifers by the million enter the blessed halls of motherhood for the first time, and this past month and a half has been no different.

Being the next best thing to a hired man, I am promoted each spring to this position of dubious honor, the pay is no heck, the feelings of satisfaction are worth it once I get myself pulled back together.

One thing you can say, it's exciting. Several times I've been conscripted into carrying a new calf into the warmth of the barn - an endeavor I always get a lift out of and sometimes two. Here I've saved a wee creature from the elements - wonderful! The second lift, not anticipated, is when the mother lifts my 5 foot 4 frame seven feet off the ground in her anxiety over the new baby!

No time is sacred during calving period. Anytime, day or night I've been on call. At 4 a.m., ten minutes to church time or just when I've popped a fussy cake in my cantankerous oven.

One thing I'll say for it - it makes a wonderful conversation piece! Imagine greeting your minister with this remark: "Sorry I'm late, Reverend. Had to pull a calf. Whopper it was too."

With a call from the Chief Surgeon, "Come on, Bets", I'm off and running, my old number 10 rubber boots flapping as I head toward the outdoor maternity ward...

It's the night calls that are particularly exhilarating. I'm a born scaredy-cat. I can almost feel myself being gored to death against the old corral, by the new mommies protecting their young from predators. So much so, I almost ride in Bill's back pocket while crossing through the herd. Actually, if one starts to bawl and paw the dirt, I'm off and running, glopping through the spring goo. The only time I hold my ground is when my boots submerge. Then my yells scatter the herd through the whole pasture.

Still and all, while I'm assisting Bill, handing him the thingabobs from the creolin pail, I feel like a real Florence Nightingale with a dull flashlight.

Guess what I got for Easter last year? Roses? Perfume? Nope. Two bright shiny dooeys for pulling calves, all made of shiny chain. Isn't that romantic? They do work a lot better than the old-style baler twine though, and after all it's the thought that counts - if it was perfume, the smell of the creoline would cover it no matter how strong.

Also in the line of duty, every year or so, we get little orphans or slow pokes. Feeding those takes more talent than putting on a comic opera, and the odd time it almost seems to turn into one. You sit on an old tub, prop up the pop bottle with one hand, pry the mouth open with the other and just as you think you've got everything under control, down come the teeth, crunch, on your thumb. It's enough to send a healthy girl back to homestead in a mighty snit!

And let's not forget those miserable wet beings who are ushered into this world, plop in a mud puddle. It's off to the bunk house with them, if you're lucky, or your kitchen if you're not, to be brought back to life and warmth... always remembering that they are supposed to be worth more on foot than you or your floor.

Ah yes, all these adventures mixed in with scours, ear tagging, creolin and the smell of stew, have earned me not only my diploma but my Masters, I'd say!

Sad Irons Were Sad!

Guess what I heard the other day? Sad irons are selling for as much as fifty dollars at auctions!

I'm in the dark about this fact, because when I was a kid every second housewife would pay you ten dollars to take it away! Only then she wouldn't have anything to iron with, and clothes in those days had to be ironed - organdy, cotton and voile had to be ironed 'til they stood up by themselves.

Starch was used by the bucketful and every last item, except socks and underwear, was starched each washday, and they had to be ironed - it was the law of the west.

So our mothers were forced to maintain a truce with their old sad irons. This was a real test at times, as they were such miserable things to use.

First of all, you had to get the old coal stove going until it was red hot. Then and only then would the irons condescend to heat properly. This took at least a good hour and in the summer it heated the kitchen 'til it was like a sauna.

If you happened to be one of the fortunate elite you probably owned a gas iron, but most of us had an assortment of the other kind.

Some had their handles intact and they were terrible. They had to be held with a pot holder and two dish towels and you felt like you were ironing with boxing gloves on.

The others had a removable handle which you clipped on each iron as the last cooled off.

They all worked about the same - like a cross between a box and a brick. You had to hook your handle on and run like a devil to iron a garment before the iron cooled off. This was rather disappointing as it took it yet another hour to heat again.

But it could lose what heat it had if you as much as sneezed enroute from stove to table. Then you had to iron with unbridled fury in order to get your ironing finished by noon.

Another problem was the streaking. You never started ironing without first rubbing the iron about five times over a piece of brown paper, and even then you could send a mighty streak across whatever garment you were ironing at the time. Usually this was a frilly eyelet dress which you had to wash and dry again. Of course if you stuck the iron over the sink it would cool and off you'd go again.

It also had the habit of scorching things. It seemed as though those irons were either too hot or too cold most of the time. If you weren't careful you could take the back out of a blouse in one swoop and have to set it in the mending basket if there was enough left of the garment.

Our iron had the built-in handle, and it was some time before we got two of the other kind. It was terrible. Either you were burning a hole in something, or else it was so cold it wouldn't iron a Kleenex. You felt as though you were ironing an elephant's hide as the bottom was so scored. By the time we got the ones with the removable handles we felt like the Rockefellers, even though the handle would let go once in a while and land on the cat.

Of course by the time we got an electric iron we held it in the same reverence as MacKenzie King. Yes, the old sad irons might be worth fifty dollars now, just as conversation pieces - they did play a big part in the settling of the West!

Boy It's Cold!

In today's world of central heating and electric blankets our minds rarely weave back to thoughts of by-gone times and cold houses.

The day of the 'cold feet' joke has about had it, and the kids of today would stare at you in disbelief if you mentioned how you could get dressed in three and a half seconds on a cold morning.

Heavens, we're badly done by if the power goes off for half an hour. It's a soft, warm life - warm rooms, beds, fireplaces in each corner. There's no difference between night warmth or day warmth but we all take it so much for granted.

But boy, the stories of by-gone days. They haven't been stretched either. They are incredible enough as is.

Winter time meant the "Mighty Change". Every kid started producing sort of a human anti-freeze come October, much the same as rabbits turning white.

All the parents in the country got out the big bottle of cod liver oil and it was duly administered each morning.

Rugs were hooked and quilts made but even with these aids it was a real feat of courage going to bed and, getting up the next morning was an even bigger one. Even with the heater blasting away in the front room and the old cook stove red-hot in the kitchen, the bedroom floors still remained an icy cold. So before bedtime, everyone crowded around the heater and waited 'til they started to sizzle before taking a mad run for the bedrooms.

The quilts were regarded with the same pride and respect as Dad's best saddle horse and they were bartered over in like manner. "I'll trade you my three mouse skins for two nights of the orange quilt!"

I don't think anyone actually suffered too much - mainly

because there were so many kids in one bed. They can talk all they want about the cause of the large families in the old days - I can tell you without a doubt. It wasn't the absence of the PILL. It was a survival tactic. Parents knew if you wanted a warm family you had to have lots of kids. It was a simple law of survival.

The mornings were really something. Even though Dad had got up an hour before to shake up the fires, you still never got really warmed though until about noon, even with a hot bowl of porridge in your stomach. Even your chilbains didn't start itching 'til then.

You could put your underwear between your mattress and the bottom sheet but you still had to bare a bit to get dressed. It didn't take long to learn that if you hopped on one foot while dressing you would have one warm foot and if you hopped quick enough you wouldn't stick to the linoleum.

Mothers were always yelling at you to stay back from the walls because the linoleum would always curl up six inches from the wall on a cold night and you could break a chunk off just with a tap.

The school wasn't much better. The stove had to heat up there too. The kids huddled together so close it left imprints on everyone's skin. I guess that's why so many kids married at sixteen. You can't stand that close to each other for six months and not start up some chemistry!

Old Wash Days

Monday - wash - Tuesday - iron - so on. Remember that old rhyme? Monday blues have been talked about for centuries and for good reason. It's bad enough having to do the wash without having to be reminded of it by a nursery rhyme.

Even as a kid I hated wash day. One good healthy reason for this hatred: I was drafted into hauling the water from the creamery. I had a brother but he was smarter than I and would always beat a retreat to the lumber yard come Monday - he finally got a job there!

I found out years later he used to sit in the office and read funnies while I slaved away to keep the family name unmarred and the linen clean. After all, cleanliness is next to godliness (that's another saying I've learned to hate).

Pail after pail, two by two, I hauled that wash water. I felt a bit like a modern day Robinson Crusoe who was trying to empty a sea with a tablespoon. To this day I'm positive hauling all that water was the reason my hands hang down to my knees.

After I got the boiler filled I started on the wash tubs. When I finally got everything full and my shoes emptied of the spill-over, it was time to throw it all out again.

Washdays didn't improve in popularity when I got married either. Washing was something all my neighbors had contests over to see who got theirs on the line first. All except me: I had a private one to see how long I could put off doing it at all.

When the day finally rolled around (when the dirty clothes hamper popped its lid), I'd be forced into action. I'd sort the clothes but the pockets gave me trouble - I'd invariably miss something. Bill's wallet was the main thing. He had the cleanest money in the district until he learned to check his pockets himself.

Winter washdays added fuel to my hatred. Hanging clothes out to freeze solid or hanging them in the kitchen was the pits. I think it was due to the fact Bill got swatted in the face with frozen clothes so many times that I got a dryer when I did.

Another thing, I always marveled at the friends who took pride in hanging their clothes on the line just so - a pair of socks, all the towels, matched pillowcases. They did look lovely, but they didn't dry one bit faster than mine - a sock, some underwear, etc.

Times have changed, though. I still don't care too much for washing, but it's so much easier now what with all these mechanical wonders. Why can't a woman have these things when she's got five kids at home and a hired man?

Winter Styles — How They've Changed

As I view our younger generation today, I see them much the same as we were - two legs, big grins and the world by the tail.

But there is one big difference, especially in the cold weather. All you see are bright little eyes, feet in boots that resemble those worn for space travel and not a piece of skin showing anywhere.

I can't help but wonder how much time it takes a Mom to get a kid dressed on an Alberta winter day.

I heard the other day that the younger ladies in Weight Watchers fail to show up in the winter - small wonder, those moms burn off more calories in the winter months than all the flare pits in Alberta.

Now when I was a kid it was different. I was a war-time kid. That meant no father to correct you, meat rationing and a shortage of winter clothing, the latter causing you to lose any fussiness which might lurk in your subconscious.

I'm sure our generation will go down in history as the "Bare and Frosted" years!

The clothes that our parents purchased had to last. Mothers had the unique skill of picking clothes for durability and warmth - how ghastly.

As a girl I'd rather show a bit of frostbite, no matter how painful, than seven inches of blue bloomers! But that's what we had to wear. Big old baggy, blue bloomers. Horrible things. Whoever invented them must have had a grudge against all womanhood.

I tried bleaching mine one day when Mom was out. I even found some old lace in a drawer to finish them off. Actually there was very little left to finish off when the bleach got through with them. I had doubled the quantity, and what I lifted out of the water vaguely resembled a blue tinged Curly Kate pot cleaner.

Besides catching the dickens from Mom I also got another pair - bigger and baggier than the first.

But when we hit the teen years things were different. In fact I'm sure all of us girls must have had a guardian angel watching over us in the winter - why else do we still have lungs, limbs and ear lobes?

We didn't have fine sealskin boots. Heck no. Remember what boots we could get were very inferior, and did little to keep our feet warm. We used to wear two pair of socks in them and bare legs. Of course you could purchase cotton stockings but we wouldn't be caught dead in them. The only girl who wore them was an elevator agent's daughter and she was unlucky enough to be warm from head to toe - what a disgrace!

The chronic complaint of all kids was chilblains. Now for anyone not knowing what chilblains are, let me explain. Take the itchiest mosquito bite you ever had, spread it evenly over your toes and the sides of your feet and set it on fire and you've got a mild case of chilbains.

Grade seven, eight and nine kids spent each winter day sitting at their desks writing with one hand and scratching with the other.

It's strange we thought it rather sexy to show off wind-chapped knee-caps. Of course the high school girls had nylons and these they flaunted - even after they had a run or two which they stopped by dabbing on red nail polish.

We were a healthy lot, though. Our recesses were spent outside even in the severest of weather. We had goose bumps galore but no colds, possibly because our parents tossed cod liver oil down our throats from July 'til June Whether it actually worked its proverbial wonders or not, I'm not sure. One thing it did, though - it equallized us. We all smelled the same - like stale cod fish!

Winter Fun

As I sit here by my window, I view people of all ages zooming over crusted snow on ski-doos, whipping up the hills like greased lightning, their occupants just a blur of colors as they pass.

It's fun, to be sure, and ski-dooing is certainly well established as Canada's number one winter sport. But I can't help but wonder whether it has an edge over other winter sports, especially those of by-gone years.

You know, I doubt it. Thoughts of winter fun when I was a kid clammor forth and these are of pure, unadulterated joy!

True we certainly weren't motorized. We couldn't reach the speeds of fifty or sixty miles an hour and we didn't look nearly as classy as our counterparts today but we didn't need gas - we did have a gas though!

Absolutely the most fun was to saddle up our friend's saddle horse and toss a rope on the saddle horn and off we'd go - one in the saddle another on a pair of battered old skis. These affairs would break the heart of a ski instructor in today's fashionable world, but they worked and you could wear any boot you owned. Just adjust the leather strap. Anyone from size one to eleven could wear them.

The world fairly whirled as you flew by, sometimes straight, sometimes off to the side. You did have to use a bit of caution, or you'd end up running into the back of the horse. When this happened the horse never seemed to kick - I guess he figured we might tie ourselves to some other part of him.

Greenpeace had nothing on some members of the community back then. What dreadful children - tormenting a poor animal - they formed a group but they had no money so nothing came of it.

We also had real old-fashioned sleighing parties on every slope within miles whenever we could get together.

All kids from every walk of life would gather together with everything from homemade sleighs to battered-up toboggans. Our get-ups were colorful, if not stylish, but style mattered little. Dad's long underwear, hand-knit sweaters, big brother's socks and an assortment of homemade mitts and scarves. Some fit and some didn't but who cared?

The odd rich kid showed up with a fancy new store-bought scarf but the poor thing was too busy untieing it from fence posts or yanking it from under sleigh runners to flaunt it. We were a clan and as a clan we felt we should all be the same ... poor! Every mode of travel we shared. If there were a pair of skis they were divided - one per kid. Just try going down a hill standing on one ski! That takes some balance.

Washing machine lids were terrific. Anyone lucky enough to sneak one of those from under Mom's nose was a real hero. These treasures have a two-fold ride. If you started them off right they would spin madly around as you shot down the hill. Of course the odd mother would spoil all the fun when she came shrieking along to save her precious laundry unit while it still had some enamel left.

Another favorite was the cardboard slide. These we made by splitting a box open and waxing the bottom thoroughly. Of course this presented another problem: where to get the wax. We'd save our pennies like misers to buy the odd can, and our poor mothers had to resort to hiding their tins because we could shoot a tin a Saturday. My mom hid hers in the coal bin because she knew full well we'd never find it there - I had such an aversion to hauling in the coal.

Wash tubs were fun if you were daring, but they had a nasty habit of digging in and sending you ten feet out in front straight on your head.

We had several hills to choose from, some quite gentle and others real sizzlers. We always managed to con some poor little kid sister or brother into coming with us - not that we liked their juvenile company but rather to use as guinea pigs. If the hill looked doubtful we'd perch the poor little kid on a sleigh and shove him off, screaming all the way. If he made it safely to the bottom we tried it but if he hit a tree, got tangled in wire or went over a ledge we'd go down, pick him up, wipe his nose and warn him if he tattled on us we'd send him down again. Needless to say we needed a different kid each time - one year we went through twenty-four.

Old Daredevil cliff was our one real challenge. Steep,

winding and rocky. Anyone brave enough to test-run it was a real hero.

After our outings we'd head home, soaking wet, bone weary but oh so content. The supper awaiting us tasted like a feast - even turnips tasted great.

Yes, I'll place my trip on a ski behind the old horse and that trip down old Daredevil cliff up against any ski-doo party of today and I'd love to see a kid today try going down the coulee on one warped ski! That, my friends, is a challenge!

Christmas Long Ago

Christmas season is magical. Even in today's world of hustle and bustle there's still a special feeling one gets only at this time of year.

But somehow the Christmases of long ago were extra special. Why I'm not sure. The pioneers certainly didn't have as much material wealth as we have today. They lived in sod houses or plain ones made of lumber hauled by a team of horses from centres fifty or sixty miles away. Money was short - but still Christmas was extra special.

The mothers would start preparing way early in the fall, knitting away to beat the dickens. Scarves, sweaters, mitts, many from wool from their own sheep which they carded and spun themselves. Many times she'd hide her work behind the wood box or in the washtub when she heard steps approaching.

She made her own mincemeat, and her plum puddings which contained potatoes and carrots but tasted as good as any English plum pudding to be had back in the old country.

Christmas cakes were baked in October and if she managed to sneak a cup of Dad's brandy they tasted like heaven too.

The Dads also were busy. After chores were done and the kids in bed they'd make toys from odd bits of lumber. They'd saw and hammer, paint and sand, and what they turned out might not have a motor, but in the kids' eyes it could climb Mount Everest.

The sleighs turned out were sturdy and lasted for years. They could haul anything - from the baby to seven kids. The runners were smooth and the kids could sail down the hills and half way up the next one.

At school the teachers practically quit teaching the ABCs come November, for then it was time to start working on the Christmas concert. The excitement when the parts were given out was almost unbearable. All the girls wanted the biggest parts and most of the boys did too, but they couldn't admit it so they'd groan loudly and pretend it was a dreadful bore.

The night of the concert was almost as big a deal as Christmas itself. The girls had their hair curled in five thousand ringlets and they blossomed out in new dresses, made from everything from a bleached sugar sack to the skirt from Mom's wedding gown. But they all had one thing in common - they all seemed beautiful that night. The boys were too, as were the stage, the Christmas tree and the colored candles. Every kid got a gift and a bag of candy which they ate as they drove home in the cutter after the concert.

In December there wasn't a day that the kitchen didn't smell like heaven. Cookies all decorated, tarts made from mom's mincemeat. The best part for all kids was snitching the cherries and raisins while Mom was baking.

The Christmas tree was terribly important. About two weeks before the 25th, all kids would bundle up and head off with Dad, pulling the sleigh behind them; the forest wasn't far, and spruce and pine trees grew aplenty. But the choice was so difficult it usually took up to half a morning. One was too big, another too small and another too cute to cut down. When the big decision was finally made Dad would chop it down and they'd head home to decorate it.

I think Christmas Eve was almost sacred. And on the ranch even more so. There's something about a stable with horses and the milk cow all bedded down, eating their hay in the quietness; is it because it's so like the first Christmas Eve?

In the house all and everyone would wrap their special surprises for the rest of the kin. Then they'd heat up bricks to keep their feet warm for the sleigh ride to the neighbors or kin.

Dad would harness up the horses and off they'd go. The air would be crisp and the moon bright.

Dad would sing only one night a year, and, this was it - who cared if the odd kid was off key - no one was out of tune Christmas Eve! When they got there they'd all pile in by the big pot-bellied stove and warm up. The time flew and when everyone was stuffed on whatever was traditional in that household and the gifts exchanged, they'd head home so content.

The Mom wore her best gown covered with a gigantic apron and Dad had his new sweater on. When evening arrived and the chores were all done, the family would gather around the fireplace and Dad would get the family Bible and read the Christmas story - old but ever new.

"And there were shepherds abiding in the fields keeping watch over their flocks by night ---"

Diane Wasn't The Only One!

Dating today is a quiet affair - no fuss or bother. They just dust off their jeans and off they go - both boys and girls!

But when I was a kid it was nigh on as important as Di's and Charles's dates.

For one thing, it took about a week for the boy to get up the courage to ask you and three days for you to prepare. First and foremost, it had to rain so you could wash your hair - rain water was softer, you see - no fancy rinses in those days. Then you'd curl it in ten thousand pin curls and sleep on them. When the pins came out, your hair looked a bit like an old car seat spring, but to us it was clear beauty.

Next you'd get your very best outfit and wash it and iron it - you had to watch out for too hot an iron, as it was war-time and even your very best wasn't all that great and you could whip the back out of a blouse with one swoop.

Our old saddle oxfords were polished with some stuff that looked like what you use to whitewash the fence, and the blue part we polished with floor wax. It worked, too - we should have patented the idea!

A nice hot bath was sprinkled with sweet-pea cologne left over from the Christmas concert - no Este Lauder Youth Dew for me back then! This didn't mix too well with the carbolic soap, but who cared as long as we smelled?

Of course, any time I had a date I was left open for teasing from my brother.

"Boy, he sure must be hard up for a date asking you out!"

"One thing, you won't be bothered with mosquitoes, smelling like that -- gasp, gasp!"

But even these remarks I dismissed with the thought - well, I was the one he asked out, so there!

Getting on towards the great moment I'd be combing, dressing and applying great gobs of bright red lipstick which was all the style then and which you could almost start a fire with.

When the knock came at the door I almost flew to it in order to beat my brother, and I'd lead the fellow in and leave him with Dad for a moment - I knew you should always keep them waiting a moment no matter how anxious you were to get away. Dad of course would try to discuss the politics of the day.

"Well Bob, what do you think of MacKenzie King?"

"Gee sir, I don't know anyone by that name - maybe he goes to the Linden school!"

"Well then, what about Lester Pearson?"

"I don't think I've met that guy either!"

Before Dad gave up completely on our generation I'd hurry my date out.

All us girls used to wish for hand-holding and sweet nothings, but all I got in the way of whispers was:

"Hey, who the heck is MacKenzie King?"

So You're To Marry A Farmer!

More and more of our young girls are sporting diamonds, and more than a few of them have chosen farm boys to settle down with. I, for one, am very happy to see this, for farming is certainly a challenging occupation!

However, I feel duty-bound to pass on a few tips which may help them in their new venture, having learned by trial and error myself!

First things first. Pick your wedding date with the utmost care, especially if you're the sentimental type. For anniversaries July is out, because then you'll be busy hauling cold lemonade out to very hot men in the hay field.

September, October and even November also are taboo - you'll be harvesting and nothing but nothing interferes with harvesting operations. At that time you'll find yourself chief cook and bottle washer, chore boy and errand runner, all mixed in with the mobile restaurant you'll be operating in all corners of the farm.

The safest time to plan your wedding for would be the last five days of June, between the spraying and the haying. Having your anniversary at this time would ensure time for your better half to take you to a picture show or an auction to celebrate the occasion. He might even let you buy a new milk cow! Now, isn't that what you've always wanted?

Next, give up on the five-day week with Saturday for togetherness. It just doesn't work that way. Your farmer will work all day every day and half the night in good weather and then, come a rain, he'll sleep for twenty-four hours with scarcely a snore.

Rain? Oh yes, rain is different on the farm. In the city you have spring showers, thunder showers and sprinkles. But here on the farm you get five tenths, seven tenths, drought and that million dollar rain. Now some day you may wish to bring on a good rain and this is easily accomplished by leaving your clothes on the line overnight, scrubbing your porch floor or washing the family car. To break a severe drought you may have to resort to all three.

Now toss your routine out the window. You need a very

flexible one out on the farm. No eight to five working hours, with meals at seven and twelve and dinner by candlelight. The only time you'll use candles will be when the power's off and then you'll wish it was on! You can usually manage breakfast okay but from then on, anything goes! He may be in at five when he said seven, or three when he said twelve or vice versa. The closest you'll come to pinning him down will be something like this, "If the hitch don't break and I finish the fifteen acres, I'll be in between four and nine!"

Also, you may be your own twosome for meals or you may just have six or seven heads pop in along with your husband. So always be prepared. Don't worry about gourmet cooking, either. I firmly believe that you could feed them a bowl of chop and they wouldn't notice, being so busy talking over the wheat situation, that new malting barley or the line-back bull so-and-so bought!

Don't expect to get everything you want or need. Money is used in strange ways here on the farm. The code of ethics is will it pay for itself? A combine will, a bull and six heifers will. But a rug, a sofa, or even a pound of cheese is strictly luxury and you must wait!

Always look on the car seat before sitting down. Three to one you'll find a dirty pair of overalls, a grease gun and three parts for the tiller on the front seat and maybe even a calf on the back seat if the local vet is to be visited.

Be prepared to be a built-in bookkeeper, calf sitter, and chore girl all wrapped into one bundle right along with housewife, lover, and mom.

However, don't let me scare you off girls. You'll love it!

Now For Some Directions!

I just can't understand it! How on earth did that crop inspector end up at the town dump?

I distinctly remember giving him very specific directions to that wheat field! I told him plainly to go past the petunias, out the lane, past the white house with the blonde wife, turn east at the place with a well in the backyard, keep going til you come to that yellow or green (I forget which) playhouse in the yard, turn west and head by the farm which owns that floppy-eared dog. Then go four miles north. You can't miss it!

I always run into some snag or another when it comes to directions and this I have to do twice every week, it seems.

There forever seems to be somebody searching for Bill, and I try my level best to steer them on the right course.

"Oh, Bill's over at his VLA land. Go out our west lane because the south one is too bumpy. Head south over the railway track, pass the garden which needs weeding (next week with luck), over the track again, past the neighbor's horses, over the track once more and turn left. If he isn't there come back one eighth of a mile and go under the track!"

Naturally when the poor fellow gets there, Bill has finished that piece and is moved to the MacKenzie place. So the muddled-up man comes back for some more of my explicit directions which I'm only too glad to give. It isn't 'til Bill comes home that I learn that he was a salesman representing a compass company.

Part of my trouble is I'm never really sure where either Bill or the man is. If he says he's going to hay west of the house, then chances to one I'll find him down at the preemption cultivating, and if he states he's going down and see how the oats are doing, I'll automatically know he's stopped at old Fred's to see how the new bull is making out.

Then I also get into a bit of trouble when I send Bill to town for groceries, like a tin of honey.

"It's in the second aisle as you come in the door, past the tea bags, east of the cake mixes, west of the floor wax, beside the peanut butter. Get the large tin on the top shelf behind the gooseberry jam."

Usually by the time he gets to where I sent him he's forgotten what he was sent for and comes home with three pot cleaners and a tin of snoose for the hired man!

Of course there's always a little bit of excitement when I send the kids to the garden for vegetables.

"Get me some of my early potatoes. They're down the first row on your left with deep purple blossoms, and some leaf lettuce and a sprig of parsley."

Something like two hours later, in they wander! With fifteen pea-size potatoes, some cabbage leaves and a dash of parsnip tops.

Actually, about the only time I ever got anyone to where they wanted to go was when that good-looking fellow with the briefcase wanted Bill on the sixteen acres. How was I to know he was the income tax inspector?

Oh well, it stands to reason that anyone who gets lost in the Bay might just have a wee problem with their sense of direction. I've probably used up all my sense trying to figure out why I'm on a farm in the first place!

Please Take A Bone

To get an otherwise sane man ready for a hunting trip is just about as hard as defecting from Red China. You see, men didn't marry us just to gaze into our baby blue eyes or peek at our shin bones, no sir - they married us for our little helping hands, willing or not.

While he loads the food for the horses, you pack up the food for him! Apple pie, roast beef and those dreadful oatmeal squares which stick to the ribs for four hours and then expand. Then you pack the Tums in case they stick a bit longer than usual.

Next you drag the old woolen underwear out of moth balls and hunt up the red hunting shirt. Then you lay the old navy kit bag out on the bed and start tossing stuff in while he reads off the list.

Guns, ammunition, the saddle horse and the grey pack horse are all loaded and He's off and away. By the time he pulls out of the lane you have to quell the urge to bar the door in case he comes back!

Running the ranch is comparatively easy - so what if the hydrants freeze, the bulls go off their feed and the milk cow pops foot rot? Marriage to a farmer isn't supposed to be a bed of roses anyway.

Actually, it gives one a sense of elation, running a whole spread by oneself. You feel somewhat like a team from Little House on the Prairie, only with running water!

Just as you're really getting into the swing of things, back comes the mighty hunter. Over a cup of coffee he tells and retells the story of the hunt. You're almost afraid to ask him if it was successful - probably because your idea of a successful hunt is competely opposite to his!

"It sure was a dandy, Bets! You should see the rack that moose was carrying!"

You don't tell him you wish it was still carrying it somewhere out by the James River - you have to be a bit of a sport, now don't you?

Early next morning finds you out in the bunk house laying paper everywhere and kneedeep in moosemeat. As you wrap each piece you wonder just where you'll tuck it in the deep freeze. He whistles as he saws away - resorting finally to the buzz saw. Everything from a chunk of the neck to a piece shaped like a rubber boot is blithely called a rump roast. The odd bits and pieces are hauled to town to be made into hamburger, and you wish in vain for the neighbors' dogs to invade the bunk house before it's too late. But no luck - into the freezer it all goes.

As the months wear on, so do the jaws from chewing and the dog just howls and runs away when you toss him yet another chunk. Of course while everyone is chewing bravely away, the cause of it all relishes in telling yet again, how he outwitted stalked and bagged the game - omitting the little fact that the creature was 90 years old and too tough for the wolves to bother with.

Putting it through the meat grinder is about the only way to tenderize it, and this is plain hard work. Actually, about the only tender thing to go through the gears is part of your thumb.

Besides losing the odd thumb, you also notice your friends staying away; becoming none too fond of all the little gifts you've been easing on them each visit.

Finally you end up sneaking the last ten packages to the dump but just as you breathe a great sigh of relief with visions of a nice roast of pork dancing around in your head, you hear the terrible words - "Boy, just three months 'til hunting season - I'd better go to get us some good meat for the winter!"

Hair — The Crowning Glory

I marvel as I wander through our drug stores today and note all the hair care products to be had. There's so many shampoos and rinses it's very hard to make a choice. There's perms and tints and straighteners and creams and pomades and mousses.

But still to me, hair can be a blamed nuisance. It's always dirty when I need it clean, and straight when I want a curl or two. It's neither a complete color nor a lovely grey. Even when I was a kid it gave me trouble. First, you see, it had to be washed. I had a very thick head of hair which needed washing often, and before I could do this it had to rain. Nobody, not even the dirty kid down the street washed their hair in town water. This was an unwritten law. Why I'm not sure. The town water was soft enough and certainly plentiful, but rain water it had to be. Then it had to be heated in a pot on the old cook stove. The tea kettle was never allowed to be used in case one drop stayed in it and spoiled Mom's cup of strong Irish tea - although I'm sure nothing could taint Mom's tea, it was so strong!

As far as shampoo was concerned we rarely had any. We certainly didn't have the variety seen today. Poor stupid kids we were. Missing so much. We didn't even know we were missing love and such, popularity and dandruff because we didn't use shampoo. Instead we resorted to Lifebuoy soap and a shot of vinegar in the rinse water.

Actually we had very little dandruff but we did smell rather peculiar - rather like a dill pickle. It didn't matter, though, because we all smelled the same except for the banker's daughter, who was fortunate enough to buy Watkins Coconut Oil Shampoo which her mom got in 48-ounce bottles. She used to sell an ounce or two to us less fortunates the night before our school Christmas dance.

Of course we had to curl our hair (to use the term loosely). This we accomplished in various ways: by winding five hundred pin curls and pinning them down securely with bobby pins, by using real fine metal curlers which left permanent dents in our skulls each night, or by braiding our hair wet and leaving it until it was dry.

The results of all these methods were basically the same. By the time we got it combed out we looked a bit like a cross between an Australian bushman and a raggy floor mop. We thought we were beautiful, though - the bushier the better.

At recess we'd comb each other's hair in various styles. Back combing was still to be discovered but we didn't need it - we actually had a back-combed style long before it was ever used in the fine salons of the world.

One thing, we didn't have to curl our hair every night like some girls do today. Ours was so tightly curled it took a good week to straighten out. Another nice thing, we had short hair the first of the week, medium for the middle and long straight hair for the weekend date!

There may have been wigs back then but most of us had never heard of them. The only tint was peroxide and that we couldn't afford.

Why none of us didn't go bald what with the tight curling, the harsh soap and the vinegar rinse I'll never know. Any of my old school chums that I've seen still have all their hair, so maybe we had something in the way of hair care back then!

Riding Herd

Our cattle pasture in the summer months is in the forest reserve west of Sundre. It's a beautiful spot. The cattle and horses are belly deep in the succulent green grass. You can ride through the forest guiding them to higher range and can spend a blissful few days away from it all. But to see all this beauty you have to ride a horse!

One year I was particularly glad to see the cattle back home. You see, I for some inane reason decided I needed a new adventure, and thought I'd help the cowboys round up the cattle so we could bring them home via cattle liner.

I really looked forward to it. And Bill, being a man short, thought he might have a job for me even though he knew I was a bit short on riding knowledge, to say the very least.

I knew I was straight-A dude from running shoes up, but I did ride a bit and loved it out west. So before I could be overtaken by what little commonsense I possess, I was out there, standing by my trusty mount, and sporting an old pair of chaps which hung 20 inches beneath me. I was full of ginger and spice, fresh lipstick and raring to go.

My little horse, Dusty, knew me so well. We had got to know each other during the summer and each had a healthy respect for the other - Dusty not expecting too much from his rider.

As we rode along the clearing I felt like a real cow-hand. The valleys below were a mass of autumn gold and rusty orange and I soaked in the beauty like a wet sponge.

Soon we came on the main herd. And from there on it was utter panic, pouts and sore spots for me. What on earth John Ware saw in riding and ranching I'll never know.

Strange creatures, cows. Instead of being glad to see us and delighted with thoughts of home, they hit forty ways at once. I gathered this was par for the course, and the men weren't unduly alarmed. They just hit off through the trees, whistling away. I, meanwhile, was trying to nudge Dusty into even a slow trot. I didn't succeed.

I got a little branch to encourage him with, but it broke at first whack. After much screaming on my part and a bit of horse sense on his he decided if he was ever going to get rid of me he'd just have to go along with it!

Bill let out a shout, "Stop them at the seismic line, Bets." And I did - all but 89 of them. So we had to collect them all over again. But now I was getting the feel of it all - in more ways than one. We got the herd moving nicely and Dusty kept up with the herd. But what with slow horses, sore bottoms and flies, the glamour started to wear off.

One thing I learned quickly. There just isn't room for a horse and two legs between two jack pines four inches apart. I know - I tried it several times. I found myself speadeagled across the valley with pieces of hide on ten thousand trees.

Bill kept yelling at me to make the horse obey and go around such spots, but Dusty was having so much fun! It wasn't every day he had such a creature on his back. Why, it was better than the old days when he used to stand on her feet!

Another thing about Dusty. If you spilled a cup of water in front of him he'd feel honor bound to leap forty feet in the air in order to bridge it.

Now out west there are thousands of little streams and trickles, and with each my trusty little mount would leap with the grace of Karen Kain.

By the time the day was over I was catching on. I found with the last leap there was only eight inches between me and the saddle - and I even came down on the same spot!

As we slowly moved the herd along I learned many things. Don't force your mount to go into the muskeg where he balks. Lead him over the wind-fall. Aspirins do help saddle sores!

By the end of the day I had lost my hat, some skin, and a bit of pride. It took three men to lift me out of my saddle and head me towards the cabin door. My legs felt like jelly and my head a brick. After some steak and beans and four cups of tea I felt great. Might even try it again next year.

I'll try for a different horse though - I found out later that I was given Dusty so I wouldn't ruin a good horse!

Soda Isn't New

I noticed just the other day the toothpaste companies think they've hit on a new gimmick - soda as a new ingredient for their paste. A real find - the one of the century!

Why now, all that romance you've been missing out on can now be yours if you brush your teeth with toothpaste containing this wonderful new ingredient.

Now I hate to disillusion anyone but between you and me, people have brushed their teeth with baking soda for centuries!

I know all through my youth I used a mixture of soda and salt - not because I particularly liked to, but it was all my folks could afford!

We rarely used anything else. Our mornings were made up of grouching out of bed, eating our porridge and brushing our teeth with soda and salt - then unpuckering for ten minutes.

Once in a while I'd earn a few extra pennies delivering milk for the milkman and I'd splurge my hard-earned cash on a lovely tube of real toothpaste. I'd be so enthralled I had the cap off before I got to the kitchen door.

Actually if the truth be known, it took the whole tube before I was used to the taste - it just didn't have the same zing!

I felt like a queen, though. While the tube lasted I'd smile so much I had laugh lines indented in my cheeks. But these disappeared two days after it was all gone. That's how long it took for the salt and soda to tighten me up again.

It took quite a bit of talent to measure out just the right amounts of each ingredient. If you overshot the salt you'd have a permanent pucker and if you used too much soda you felt like you were rising - like a loaf of date bread.

I wasn't the only soda user to be sure. All kids did through the war years and we didn't think we were missing out on romance. Probably because all the boys used it too. In fact we had more than one school romance bloom into marriage after grade twelve graduation.

One little problem, though - Mom was forever out of baking soda. She'd nicely get started mixing a cake and discover she was out once more. Then I'd find myself running up to the general store for yet another box.

So each time I read of this wondrous discovery I feel I should have patented the idea - that little box with the cow on the side has been brushing teeth for years.

My Old English Bike

When I view all the fancy bicycles to be got today I'm astounded. There isn't a size, a feature or a color not touched. They are wonderful - they can spin on a dime, race like the wind and are, as the manufacturers claim, the very one to beat the neighbor's kids with on the kids' social ladder.

But when I view these beauties, I realize just how far bike styles have come since I was a kid.

Looking back I recall my old bike - a great, husky, five-foot-high English model. Who the English had in mind when they constructed it I'm not sure, but it must have been a six-footer for sure. You practically had to use stilts to get on the seat.

Actually it first belonged to my brother - that is, until he felt too manly for bike riding. He was a very mercenary lad and charged me ten cents for every ride I took - five if he gave me one. If I happened to be gone for too long he'd charge me interest which I took out in hard labor, usually hauling water. That's why my arms are so long!

Still, I wanted to ride that bike so badly I'd get odd jobs for cash to use. I'm not sure what my brother used all my hard-earned cash for - probably tea to smoke out behind the school barn.

What pleasure I derived out of riding that bike! I think it was the challenge involved. The handle bars were so long I could almost tuck them under my armpits. I had to use an old applebox upended to mount it and it took real courage as well as a bit of Irish luck to ride it too. In those days we all wore flappy, wide-legged pants - the real 'in' thing of those years. With such a volume of pant leg, we had to fasten it all down or they'd get caught in the chain. If this happened there was only one thing to do - cut the pants free. For four years straight I didn't own a pair of pants which wasn't patched on the chain leg!

Some rich kids bought fancy little doo-dads to slip on the leg, but us poor kids resorted to our mothers' rubber jar rings.

After slipping it on, I'd yank up the bike which weighed more than I did and hop on and try to gather momentum to get me over the rough crab grass to the relatively smooth dirt of the back alley.

This I accomplished only about one in ten tries. Needless to say, the whole of each summer my legs and arms were bruised all shades of the rainbow, and even some colors Noah failed to see on the Ark! Once going, it was really exciting. I seemed to tower over the whole world and all the mortals - a real heady feeling!

I did have some difficulty with the corners, though. The odd time I'd turn too quickly and end up running into some unsuspecting soul, on the sidewalk. Or else I'd manoeuvre too slowly and practically drive myself into the trunk of a car.

The worst turn I ever made sent me sailing over a caragana hedge into a man's yard. The poor man had to stuff my mouth full of jelly beans to silence me. I think he was afraid someone would think he was beating a poor defenceless girl. After two days spent in bed with an honest-to-goodness broken blood vessel, I was back riding the bike again.

When I became sole owner of it, I didn't get a chance to make any money lending it like my brother did. By the time my sister was old enough to ride a bike Mom figured she couldn't stand another daughter flirting with death, so she bought one just my sister's size.

Crocus Time

One of my favorite times of the whole year is spring - when the crocuses are popping up.

I love crocuses. Our pasture can be alive with them once Mrs. Sunshine decides to warm their roots. In fact that's how our farm got its name - Crocus Coulee. Our household held Bill and I plus four girls and one boy. Bill and Ian wanted to call it Gopher Gultch but the feminine group overruled and it's Crocus Coulee.

I think anyone who loves nature must love crocuses. Is it because they bring the promise of spring?

When I was very little we lived west of Sundre and there were no crocuses there, so when we moved farther east to grain country I absolutely loved them, thinking them the most beautiful flower I'd ever seen.

First thing in the spring we'd congregate and ride our bikes out to the various pastures near town and have a real crocus hunt. We knew nothing of plastic bags then so we just packed them in our every available pocket. Even the boys picked them, although I figure now most of them used it as an excuse to come with us girls!

Some of us smartened up and carried old jam buckets and filled them with slough water to keep our treasures fresh 'til we got them home.

Then we'd sit back on the edge of the slough and eat jam sandwiches and jaw breakers.

After a bit of fun and frolic like trying to catch frogs or minnows, we'd head for home.

Then, under our ever-watchful eyes, our poor moms would have to arrange our flowers in every available dish. We'd make sure they were kept until the water smelled, then off we'd go for more. I'm sure our mothers were very glad when the crocus season was over.

It isn;t only little kids who fall under the spell of the crocus. I remember as a teen-ager walking out with the boyfriend of the day (it usually was by the day then too) and spending an hour or two amidst the flowers. Of course we'd discuss mighty issues like who was the class jerk, did I catch heck for getting in late and this sort of thing. Then we'd pick a very dainty nose gay - I ask you, how could you hold fifty-five crocuses, a jam pail of slough water and both your dream girl's hands at the same time?

But it didn't seem long 'til I had my own little ones bringing in bouquets each spring - the same thing. All lengths of stems, bits of twigs and many black bugs. I also was not allowed to throw any away, even though all five kids had picked the same day.

It just continues on - I now get them in grubby little hands from my grandchildren and it doesn't make me feel old either. In fact, I feel young as I watch each little upturned face grinning while I tuck them carefully in a vase and hand over the bugs to take home for their collection!

City Cousins and Us

Hey folks - have you noticed? Our city cousins are moving out! These days, every second family in the city either owns ten acres on the south forty, or is busy trying to find just the right parcel of land. Still others are hoping someday to own a little farm of their own.

I guess they've discovered that even though farmers and ranchers are knee-deep in taxes and dirty corrals, they still have a little extra something which all people crave - peace and contentment.

What with pavement and big cars, our city friends find they can attend to their work in the city and still hustle back to their little acreage by 5:30.

But one thing I've noticed - no matter how strong the call of the land is, they still feel compelled to bring part of the city with them. Instead of building near that grove of poplars, they have them bulldozed out and bring in 'instant trees' all packed in peat moss. Then they hire a couple of men to dig the holes, a truck to bring out special fertilizer and a city gardener with a degree to plant the trees. Of course, come six weeks later, when the magnolia trees or whatever have passed on, they are then quite happy to plant a sucker or two from your old lilac hedge.

When those ready-made rollup lawns they have so carefully laid fail to thrive in the fresh country air, the folks are ready to try out some of that special grass seed you used for pasture for the cows.

Another thing I've noticed - the city life has so indoctrinated people, they can't understand why on earth a neighbor would pop over with a loaf of fresh bread and a jar of gooseberry jam.

It also takes a good six weeks to break them into the country habit of stopping by just to sit around the kitchen table to talk over the amount of rainfall we've just had. This, it seems, is a real no-no in the city way of life.

Of course some little time is needed to ease their kids over their terror of the dog, the pigs and the old mare.

Naturally when you pull their car out of that piece of gumbo come a rainy spell, they can't understand why you refuse to take the five dollars they offer you.

And then when you have them over for Sunday dinner the good lady is thunderstruck when you toss a dish towel her way with "Come on - let's get these dishes done!" In the city you just do not enter another woman's kitchen!

Then, oh then, there's those yard lights. Thousands of them glaring over the land. Why, when Pa was through chores he'd blow out the lantern and you just weren't worth your salt if you couldn't make it to the biffy in the dark without tripping over the old barn cat.

But our new friends feel honorbound to light up hell's half acre from dark to dawn so they won't miss a thing. They soon learn that anything really worthwhile happens after dark on the farm! Besides, if they do happen to get lost, sure as shooting' someone's hired man is bound to find them on his way home from a night on the town.

Actually it takes about two years to get newcomers really 'countrified'. And then they join the rest of us farmers and sit back and watch, with hilarity, the antics of the next new farmers-to-be!

In The Ring

Horse sales and bull sales are showplaces, not just for animals but also for the people. I've been to both on many occasions and each time I learn something.

There are several types of people who can be identified by looks and manner if you watch closely.

Females first. Any ladies with their hair fresh from the beauty parlor and nails long and polished with a wisp of Chanel No 5 wasping about them, are on-lookers only. They're mainly out for a high time, or seeking a likely-looking rich rancher for a husband.

Giddy, cheerful girls chatting with all and everyone are generally housewives who are there simply because they would go just about anywhere to get off the farm for the day. They've usually got a touch of cabin fever and if they don't get out somewhere they might get desperate enough to mend the blue jeans or clean out the back porch.

Family groups are out in abundance and kids are clinging to Mom's hand with one hand and rubbing their chocolate bars into her jacket with the other. The look on Mom's face should be read as follows: "That damn teacher didn't know what he was saying when he told me to take Dennis to the sale to learn a bit of horse sense - next year I'll send the kid with HIM!"

The wives of the men selling that afternoon are easy to pick out. They have bits of straw sticking out of their run-down boots and a face nigh on expressionless. They are trying desperately to look nonchalant, not daring to tell a soul that their summer's groceries (not to mention wardrobe) is tied up completely in those beautiful beasts out back. Their nails are broken and they've a bruise on one elbow which they acquired while helping load the stock this morning.

You'll notice they hold their purses in an odd way - that's from carrying the pitch fork for so many months, a habit that will wear off by July.

The men are out in great form and are just as easily picked out. The ones in new western suits, waving to one and all and shaking hands with each other are the bid takers and public relations men. They have to convince you they are bosom pals so they can work you over later in the sale.

The auctioneers are dapper chaps. Rather citified possibly - they have also been known to drink from pop bottles, the contents of which tend to cause hiccups, and a slight stammer towards the end of the sale.

The ranchers who aren't selling any stock but have lots at home are sneaking around trying to convince you there's not a decent animal in the barns, but you should see the beauties on his farm ten miles west.

Out back in the barn you'll meet the true masters of the day - polishing, currying, and talking - as well as serving drinks from the old bull box to likely customers. They are dressed in down-filled vests and dirty jeans, but they are the heroes.

The buyers march about with great pomposity, smartly dressed and carrying clip boards with important frowns on their faces - but don't be fooled - these buyers are bluffs. They are usually part-time farmers with about ten cows and one old horse for the kids to ride on week-ends. The sale gives them a chance to feel important at least once a year.

But cast your eyes about. The ones to watch are the old ranchers with battered hats and a wad of tobacco stuck in their jaw, walking slow, talking little. These are the real bidders - so cautious in bidding you can't see a muscle move, but they've got the know-how and the big money.

As for the animals - if one lies down in the ring and sleeps, he's a real heller and has been tranquilized out of his senses!

My Old Skates

I went down to the basement today and guess what I found? My old skates. I could have cried. They were all moldy and I burnt them!

Don't laugh, those skates went back to my seventh grade. To earn the money to buy them I delivered milk all over town for the huge sum of 25 cents a night. The skates cost $9.95 so you'll realize how long it took me. I must have started saving for them in about grade five......

When I finally brought my macaroni box full of quarters to the store and dumped it all out on the counter I felt like Howard Hughes.

Those skates were so lovely; white with fur around the top and all; even the laces matched. The ones I had been using were my brother's old ones and they were tied with store string. As I tried them on in the store, I figured they might have to last awhile so I got them a bit big. I really overdid this, but I couldn't help but think of all the hard-earned cash that went into the purchase, so you'll understand when I tell you I had to wear three pair of Dad's army socks in them the first year.

I loved skating, possibly because it was just about all we had to do in the entertainment field in the winter. We had an outdoor rink in town, but most of us kids preferred the old buttermilk slough down near the creamery - this slough was made from the run-off from the butter-making process and you can imagine the smell.

But we didn't care. It was a wonderful fairyland. There were clumps of brush poking through the milky ice here and there, and these we used for many things - goal posts, trees and even mansions in our dream world.

An old barbed wire fence, partly submerged, served as our boundary line when we held mock battles each Saturday.

We spent all the winter months on this slough and as a result the very smell of rotten buttermilk hung about us. The teacher made us hang our coats and mitts down in the furnace room after three upper class kids complained we were stinking up their clothes in the coat room. They lived to regret this tattling, for we took them down and rubbed our smelly mitts over their sensitive faces! It worked too. By February they were down skating and smelling like the rest of us.

We held hockey games every chance we got, and though our equipment lacked a bit in style, we didn't mind. We felt just as professional as any Wayne Gretzky with our old broom handles and frozen buns pucks.

Oh Yes, those skates were part of my life for many years. I never did grow into them, but they did glide around the Crystal Palace in Vancouver as well as other city rinks but these auras of grandeur didn't give me near the pleasure I felt when young skating on the old buttermilk slough!

The Old Back Door

In this day and age there's not a great deal of difference between farm homes and city homes. What with electrical appliances, housing loans and so on our life styles are much the same.

But there is one difference which will never change and this is the old "Back Door" bit.

Why is it that every single person who happens to visit a farm feels compelled to pop in the back door? The only use the stately front door gets is to let in a bit of fresh air now and again. It doesn't seem to matter if the lane comes to the front, that the flowers are more elegant at the front and that the hall is free of clutter. The poor back door still gets worked overtime on a farm.

In all the days I've been at Crocus Coulee I've only once had someone come to my front door. And that happened to be a twelve-year-old boy who was lost. That time it took me ten minutes before I found where the knocking was coming from. It's a good thing he was persistent or he might still be there!

Still, to enter a farmer's back door is out-and-out dangerous. You see, there are usually rubber boots, buckets, halters, pails of nails, and any number of other things a farm collects. With five kids, some hired men and ten thousand projects always underway, ours was always "enter at your own risk!"

In rainy seasons and in spring time you can scrub the porch twice a day and still have to wallow through mud to get to the kitchen.

Also, coming in the back way you have to traipse through the kitchen to get to any other room, and farm kitchens are the busiest rooms in the house. Long years ago there could be cream cans and egg crates and even now you can find a new calf on the floor being dried off with sacks by a register.

Now a person can keep a tidy front room most times and a passable hall; the flowers are usually nicer out front and the front step is a nice smooth cement, but still it's the kitchen and the back way that's on view on a farm.

I don't mind my neighbors and friends coming in the back way

– most of them would rather sit at the kitchen table with a pot of tea anyway. But still I wonder about the new minister from Toronto, or great-aunt Mabel who lives in a $900.00 a month apartment in Edmonton. It can be downright embarrassing to watch her mince through the glops of mud and such. Then again, I think she looks on it all as sort of an adventure - like a safari in Kenya! If she would only come to the front, she could be greeted as she deserves.

I've mastered the whole situation, though I now plant my best flowers near the back door and set my planters near the porch. I've moved my brass door knocker to the back door even if it looks a bit out of place. You see if you can't beat them - join them!

The Eternal Triangle

I have to chuckle (a very self-assured chuckle) every time I read of some poor soul being one of three in the eternal triangle.

The reason for my assurance is plain. Few farmers or ranchers have time to fool around. Although the problems of rural people are many, there is little likelihood of your husband having a mistress tucked away in the city. It just wouldn't work out!

Use your imagination. Your man decides to keep a cute little bombshell in a cosy hideaway - just to liven things up a bit. Well, when on earth could he possibly visit her? Maybe the odd visit could be made in the morning when he makes the trip to town to have the chop ground, or possibly a quick hello when he takes the saddle in to be mended. But what femme fatal would welcome a bearded apparition in faded jeans with grease on his hands?

And what about those little presents? Isn't the whole idea of a 'sugar daddy' to receive token gifts of diamonds and furs? Well I don't know about your man, but I hardly think there'll be money left after paying taxes and fuel bills for too many gifts, and the only stones in a farmer's mind are the ones he has to dig out on the south forty. Let's face it, farmers are a breed unto themselves. When it comes to gift-giving, they'd consider a couple of weiner pigs or a nice sharp axe quite appropriate.

Now as far as sneaking out, I don't think he could make it. For one thing, when a farmer's wife hears he's going to town something snaps inside her and she immediately finds ten thousand things for him to do when there. A load of dry cleaning, four zippers at Woolco and some shampoo - just for a start.

Then if he actually makes it to the mistress's love nest, it would be just her luck to have her brother drop in - the one that farms west of Regina and the men would sit and talk up a storm about the right kind of fertilizer to use come spring.

Then she'd have another problem: every farmer worth his salt is a pack rat. It would only be natural he'd bring the overflow to her apartment, being as how he was paying the rent and all..

So in would come forty issues of the Hereford Digest, the old pack saddle and some weed spray. I doubt if she'd take too kindly to tripping over a case of lice spray everytime she went to the door.

And then there would be the matter of conversation. She would be used to hearing sweet little nothings about love and beauty and eyes like stars. But I doubt even a moonstruck farmer could handle that for too long. He'd want to delve into great issues like the Mulroney governent, the low Canadian dollar and the useless civil servants.

As far as gazing into her eyes by candlelight, I think he'd switch on the news broadcast to catch the weather report, don't you?

No, a farm wife has lots of worries, but infidelity isn't one of them!

Lunches — Lunches

If you bothered to ask 100 farm women what upsets their schedule each and every day, the whole hundred would automaticaly reply - LUNCHES. I agree. They not only upset your schedule, they upset your home, your ironing and your braincase.

There seems to be some sort of unwritten law followed by all good farm boys which states, "Only a wife who can congenially serve coffee five thousand times a day, take lunches to the south forty and turn up with a chocolate cake at the rural school, is worth taking home." In fact if even one of these ingredients is missing it looks like trouble, and you may just be stepping into marital quicksand.

The trouble is, when you've deceived him into thinking you dearly love taking lunches anywhere on the Northern Hemisphere, you've also talked yourself into believing it!

I'm sure you don't need to be told there's no out to this chore short of divorce, and let's face it - I'd rather make up some bologna sandwiches than do without my Bill, wouldn't you?

I've taken out, served in, whipped up, and sent over enough lunches to build the second Eiffel tower.

There's the surprise guests who arrive anywhere from 7 a.m. 'til 12 midnight, and these show up usually two hours before you've decided to bake. So you haul out your last banana bread out of the deep freeze, set the oven and keep the guests talking about the terrible drought until it's all thawed out. Then you lay a piece of cheese over the frozen centre and pretend it's a new kind of sherbet.

Of course these guests are not counted among the coffee drop-ins. All they want is a good coffee and someone to chew the fat with.

Then there's the lunches in the field. Beside every farmer is a wee woman who is able to bring out and serve a lunch big enough for an old time threshing crew and wear a pretty dress. (Funny thing - men can wear dirty jeans and a six-day stubble, but a wife is suppose to look feminine come rain, shine or high water - besides wearing a toothpaste smile even though she knows her scalloped potatoes are boiling over in the oven.)

You also need the ability to stand teasing, not yell if you sit on a thistle and not take the muffler off on your bumpy road through the field.

Another thing - you need a built-in antenna or a full tank of gas when you start out. It's only natural your betterhalf forgets to inform you he'll be moving to the other place. That's why you see so many farm wives wandering around with bewildered looks.

Of course we can't forget the social affairs lunches. There's not a woman among us who hasn't been told at ten to six she's been offered to bring a square to the meeting tonight. So she butters up a pan and pours out six cups of Rice Krispies, only to find the kids ate all the marshmallows.

My idea of heaven is sitting on a soft couch, my hands in suds while my manicurist tells me four servants are making peanut butter and banana sandwiches for the men.

Rural and Urban

My, I read some startling news the other day! Did you know there is no difference between the rural and the urban homemaker? That's what I read. What with plumbing, push-button appliances and good roads that gap - once so wide - has been bridged!

I'm awfully glad I read that article. I didn't realize we were so close to that point. But I am still just a mite dubious. It seems to me we still have a way to go. There is a difference between us and the urban homemaker. True, we partake in social events, we take our place in the community and even can be quite literate when we find the time to think ... but there is a difference. Otherwise, we wouldn't be here on the farm and the people who feel as we do, if not already on the farm, wouldn't be busy trying to save up for a spot to raise a cow or two and the odd horse.

If you're a country soul you're in the country...

The difference isn't social prestige. We aren't one notch down on the social ladder. Good heavens no! We're just lucky enough to have that extra something in our souls to handle the exciting and usually untimely events which any girl, married to a farmer, has to cope with.

You know the type of happening I mean... Any girl who has lived on the farm as long as one month will know.

Let's take an ordinary coffee party. Both rural and urban gals prepare the same way. Plug in the curling iron, press the same pricey dress and with a bit of luck have a peaceful bath. They toss a casserole in the oven, set the timer and head out.

But just as the city lady steps out on her nice clean sidewalk and into her nice clean car - a thing a country girl has never done - the country one notices the fool steers are out again and heading straight for the chop bin.

Of course, the city girl drives off peacfully listening to CBC, while poor little country gal tears back in the house, out of her dress and into her old jeans in 10 seconds flat - she's been known to accomplish this in eight but she had to reach under the bed for her sneakers! There she goes; broom swinging, steer yelling, like a streak of lightning. She automatically steps in a cow pie left just for her.

Everyone knows, when steers break loose, they get all hepped up and head for greener pastures, over the astors, through the rosebush and beside the thistle patch. The very last thing they have in mind is to lope back to a dull old corral. Needless to say the chase is a merry one, not to mention time-consuming. By the time it's accomplished the coffee-bound girl is huffing and puffing and completely wind-blown.

Now we come to the difference. Farm girls don't give up easily. So, back to the bedroom, dress on, feet wiped and shoes on. Toss the grease gun off the front seat and take off down the lane, 50 miles an hour before something else happens.

Meanwhile, back at the coffee party, the city girl sits, glancing at her polished fingernails, wondering why her country friend can never be on time! (You tell me what push-button appliance could have accomplished what this girl did with just two hands, two feet and a heck of a yell!)

Shopping brings another example. Our city girl plans a neat menu, jots down a neater list and strikes off to the supermarket.

Back at the ranch, this girl also plans a menu, hunts for the one and only pen which her husband swiped to mark down the eartag numbers and finally finds an old stub of a crayon in a rusty milk pail.

Her list is longer. She learned long ago you can't count on how many you'll be feeding - maybe just the twosome when you've cooked enough for 10.

Our girl knows full well, farmers congregate at kitchen tables to discuss the amount of rain, the new bull and the fact that Jake was smart enough to marry a school teacher!

After she's got her list ready she climbs into a clean pair of jeans and grabs her purse. Now to town. But just as she opens the door, she runs whack into her husband who brought along six friends, all dehydrated. As she serves the coffee she's informed to plan on 10 men tomorrow for meals, because he's going to sileage. Boom!! She throws away the old list and scoots to town and shoots 150 bucks just on essentials. She then proceeds to bake 12 pies, three cakes and two triple batches of cabbage rolls...

Her friend has not such excitement. She brings home two little bags of delicacies and a dinner wine and decides to have a nap while the wine is chilling.

The farm girl? She's still going strong, tossing her ironing in the deep freeze and wiping the stickiest spots on the floor.

Ah yes, there's a difference. Farm women have to be fast and flexible, and sport iron-clad nerves. I think every bride should have a trial run on a farm before saying I do. If she hacks that month she'll make any man a dandy wife!

Hey — The Roses Match!

Wallpapering is an inborn talent. Either you have it or you don't.

But let's face it - sometimes we have to do things, whether we have talent for them or not, even if the outcome is obvious before you start. But you can master it to a certain presentable point with advice and practice.

Being as how I've just had my dining area papered, I feel I can give you a few pointers.

When picking out the paper, don't just choose to match your color scheme or because you've fallen in love with the pattern. Good gracious no! Look to the future. Is it washable? Scrubbable? Will you be able to scrape mud, ketchup and chip dip off and still leave a respectable amount of paper?

Also, pick a pattern that's easy to match! Don't be taken in by some beautiful little thing which has to be matched every four feet. If you do, you'll find yourself five rolls short and your feet covered with four foot strips of unmatchables. These you'll have to piece together behind the fridge, the china cabinet or the better half's rubber boots.

Also keep in mind how well pizza, hair oil and fried egg stains will blend in with your background. Ask yourself how well it will weather the cracks which are bound to appear every time the old house decides to move its bottom, just for added excitement.

Don't figure on completing the job in two hours with fifteen minutes out for tea. Start as early as you can and hope to high heaven you'll be done in time to give yourself at least three hours sleep that night!

Kick everyone out of the house but those with a patient personality, because even the most perfect of angels will be arch enemies before the ordeal is over.

Measure, measure, measure - up and down and crossways. Along with every double roll of wall paper comes a little gremlin which automatically turns nine feet into eight and a half the very second you cut it out. Start where it won't show and work around to the areas to be seen because only the last four pieces will be as you want them - it takes that long to get the job down to a science.

Don't be fooled when you get a strip up. That means nothing. The very second you turn around it's going to flap its ugly face back down across your back, just for kicks.

Don't get too uptight if a piece doesn't actually match to perfection. Just don't tell anyone. If some perfectionist is nasty enough to point it out, just tell her you got bored with all that matching and remember, it's your home. If you want a rose with three stems and a half a bud, that's your business!

You can get your helpmate to hold a strip up with the broom, but make sure it's with the bristly end. Otherwise, you'll have unique little holes which may be a little hard to explain.

Don't worry if someone pops in. Just ask them to help.

If they are feeling helpful, they'll dig in. If they're smart, they'll get out, running the last ten steps. Either way, you'll win.

After it's all done and the mess is cleared up and your husband's foot is dried off, you can wallow in the joy of a job well done. That is if your feelings aren't all glued up with everything else in the house.

On Getting Up

I hate getting up! I hate it more than any other movement I make all day!

This "early bird gets the worm" routine just isn't for me. For one thing, my eyes aren't open wide enough to see the chairs and table, never mind a worm! (Even if I stood on one in my bare feet I'd barely notice the squish.) It takes me 'til 10 a.m. to get both eyes open - as far as activity, I'm useless half the morning. I might add, however, I'm a real going concern come 2 p.m.

I don't envy those bright creatures who pride themselves on having their wash out on the line before 8 a.m. and the house polished too. We all have our own pet talents and getting up just isn't my "cup of tea".

This sleepy-head habit of mine can be embarrassing, though. I've been caught in various undressed, unawakened predicaments so many times I've lost count. Barefoot and bleary-eyed it's easy to see I've just crawled out of the sack.

I have learned (through practice) that you can bluff a bit, however. Now, anyone who gets up at 5 a.m. and enjoys it, just turn the page. The rest of you last-minute risers read on.

One method I've used is the rake trick. As you well know, all early callers come to the back door; so with the first knock, leap out of bed, tear out the front door and grab the rake. Then slow down to a saunter and wander nonchalantly around the back. It's easy to see you've been up for hours raking the lawn (before the heat of the day, you know!)

The caller will be so surprised he'll never notice you're still in your P.J.s - that is, unless they are baby dolls. Then I'd suggest grabbing a pair of jeans on the way out. You don't want to start a scandal or anything. Don't worry about your bare feet. They are the fad this year and you'll be right in style.

If the weather is a bit too chilly for such a manoeuvre, try this one: with the first knock, shout "Come in", grab your clothes and head for the bathroom. Dress carefully, then when you hear their entrance, flush the toilet three times and come out smiling. They'll never guess! Everyone has a right to use that particular room at least once in the morning!

If your eyes refuse to co-operate, try the hair spray trick. While in the bathroom locate your hair spray and on your grand entrance hold your worst eye with your hand. Smile a painful good morning and head for the kitchen sink. The big bluff implied is that you've hit your eye with spray. Of course, the treatment is to wash it out (your guest might even help you). But for goodness sake comb your hair before leaving your retreat or it will never work!

Your could be a down-right liar and just not answer the door, but let's keep this as honest as possible, shall we?

The phone call awakening is easier to contend with. The procedure here is to lift the receiver, sneeze three times and cough twice. The sleepy voice will be completely hidden by your dreadful cold, now won't it? Works like a charm!!!

I honestly don't know why we should have to be deceitful, though. I know I do just as much work as my early bird friends - only at different times of the day. By 10 at night I'm a real whirlwind, but by then all the rest of the world is sound asleep. I just can't coincide with their hours, I guess.

If you absolutely must get up at a certain unearthly hour try this: Set your alarm clock ahead fifteen minutes. Next morning you'll naturally sleep in five minutes too long and the rush is on. Then to your joy you'll find you're ten minutes ahead! Glorious feeling, that!

Your husband will find this rather disturbing. Bill is forever saying, "The rest of the province goes by mountain standard. Why on earth can't you?"

Income Tax Blues

I believe if some smart composer tried hard enough he could write a tune called "Income Tax Blues" and win a Grammy! About March each year, a terrible virus hits every farm in Canada, and doen't ease off 'till the end of April.

I'm not sure about your home but ours at times, almost goes under with stress. True, nowadays we have an accountant do a lot of it, but there's still a lot of preparation before taking in the books. It starts with a great flourish.

"Okay, Bets, let's get at it!"

You clear off the dining room table and that's about the last you'll see of it for a good month.

"Where is my tax form?" ... and off we go.

"I never saw it, honey!"

"It came in the mail!"

After it's located (usually right where he put it for safe-keeping) the virus takes hold, its symptoms being grunts, groans and the inability to count. It slowly turns a happy home into a turmoil and the residents into shaky wrecks.

"Bets, look up all the old forms!"

I'm very delighted to find them all back to 1949 – but where's '48?

Then, of course, you need a pen. Now any home that's at all average has some unseen monster known as the Pen Eater. Any pen you ever set down is gobbled up overnight. It would be easier to get hold of crown jewels than a pen that writes. I've found them everywhere, from the old milk pail to the bread box, but at last you find one with the eartags in an old box.

Next all the papers are stacked in piles - not two wee, tidy ones - heavens no! I mean piles on everything - china cupboard, T.V., floor and six chairs.

Your life is in your hands if you try to walk through the room without touching one of these awe-inspiring receipts. You find you can breathe only upon entering and leaving this once-a-year inner sanctum.

You can't even water the plants, lest it ruffle the papers while the fellow is busy scribbling away at the table. But that's okay as the plants themselves have been given unspoken orders not to grow as much as one leaf while this statesman in a G.W.G. shirt is proceeding with his chore.

The job which, "Oh, it'll only take a couple of days, hon!" stretches into a week. You just watch the kids write their names in the dust and pray they don't date their signatures.

Naturally, as the week unfolds, so do the visitors. These ones are, of course, the ones who get their income tax done in town. The others are all home doing theirs!

When the dreadful chore is finished, once more you head to town and mail it off, meeting five friends on the post office steps who are mailing theirs.

I maintain it won't be just the economy and high taxes that send farmers off the land - it'll be all the blasted paper work!

How Men Change

I love men - we could never live without them! But a man is a changeable creature, especially when it comes to his 'before and after' marriage habits.

Before you marry them, nothing is too difficult for them; they'd walk over stones barefoot for miles for you. They'd brave the storms just to gaze into your eyes. Even at harvest time they'd shut down early just to go out on a date with you. They'd stay up 'til 2 a.m. even though they're beat tired, just to let you know how wonderful you are.

They fill your head - and heart - with all sorts of lies about how precious you are - more precious even than the four loads of wheat they could be combining if they had stayed in the field.

They bring all sorts of gifts like roses and candy; they carry your photo over their driver's license.

They warm up the car for half an hour so you don't catch cold. They'd never let you open a door, never mind the gate and they fly to your aid as soon as they see you about to pick up anything as heavy as a twenty-cent bag of peanuts.

Maybe it's a good thing they change along with the marriage vows. I doubt if I could have stood all that chivalry much longer. But we needn't worry about an overdose. Once you've settled down to real marriage, things do change.

Instead of shutting down the harvesting to run to his beloved, his beloved is out running the combine, and the only sweet nothings heard are the remarks made when you run over the gas pump.

As for the eye gazing, the only time you do this is when you're trying to get a piece of chaff out of his eye.

You can forget staying up 'til the small hours unless a heifer is due to calf, and then you might be enlisted to help bring the new baby into this old world, often at 3 a.m.

The candlelight suppers are served in the field, as you gaze at the setting sun, wondering if you turned off the oven.

And the roses and chocolates turn into things like a suckling pig or a pair of rubber boots, and by then that's really what you'd rather have.

Flowers? Yes, I still get mine - each year a bouquet of lovely wild roses are handed to me with greasy, careworn hands, and believe me I wouldn't trade them for all the orchids in the world!

Machinery and Me

"What are you going to do today?" is my perennial morning question and Bill's perennial answer is, "Don't know yet, kiddo!"

The other day his answer changed a bit. "I think I'll go and see about getting a combine."

I held back any questions I had, for I well remember a tractor he bought a few years ago. That was the time I proved without a doubt how very little I knew about machinery.

I was busy around the house when he drove it proudly into the yard atop a truck. I must admit, I didn't show too much excitement. A tractor is another piece of machinery, suitable for pulling stuff about.

Soon he and a friend came in for the inevitable cup of coffee. As they chattered away about all the machine's fine qualities, I leafed through the new catalogue.

"See the new tractor, Bets?" Bill asked.

"Yes, I did. Say, you have to run to town for some bread," I answered.

"What do you think of her?" he persisted.

"Think of who?" I asked.

"The tractor! Haven't you even looked at it yet?"

"Oh, it's a pretty color!" I answered brightly.

"Really, Betty!" (he calls me Betty only when disgusted.) As I peeked out the window he continues. "What did you expect a Minnie to be?"

"I don't know. You always seemed to favor red tractors!" I snorted.

"Didn't you even know what KIND I was getting?" He sounded dumbfounded.

"Well, not really - I knew you were getting a tractor - that's pretty good for me."

"There's different kinds of tractors, you know!" He said haughtily.

"I know, but they're all much the same. Two big wheels and two little wheels!" I protected my knowledge.

The look that passed between him and our friend was one I

knew well. It means, she cooks a good meal, but boy, she can sure be stupid at times!

When Bill finished his coffee he commanded, "Come and look her over, Bets." Being back in the "Bets" category again, I thought I'd better trot along. As I donned my coat I thought of all the little things I knew about tractors. It was very little! Oh well, I'd just have to bluff it.

"See, honey, isn't she a dandy!" he boasted.

"Sure is, Bill, but don't you think she clashes with the red barn?" I answered with bravado.

Holding back a groan he toured me around it.

"Bet it'll look nice against the summerfallow!" I tried.

Before he could say anything I thought of a goody. "How many spark plugs?" I asked with great intelligence.

"None, you ninny!"

"Oh, I thought everything had spark plugs!" By this time I knew I was losing ground.

"Does she use much gas?" I ventured.

"NONE", he gasped, as he walked around the other side.

"Well smartie, what does she run on AIR?" I snorted. Taking me by the arm he decided we should have a little talk. "Bets, it's a diesel. Diesel tractors don't use gas. They use diesel oil. They also don't have spark plugs."

By this time my feet were cold and I was tired of asking so many intelligent questions. Especially about dull old tractors.

"I'll go to town for the bread myself - white bread. It's not as nutritious as brown bread, but white bread makes better toast. But both brown bread and white bread take yeast to make it rise and a smart cook to make it - it also looks pretty on a blue plate." All that I yelled over my shoulder as I stomped off.

I mightn't know much about tractors but he knows less about bread, so there!

Trucking — If You Dare

How many farm wives find themselves behind the wheel of a truck the odd time throughout the year? A good many, I'm sure, especially now when the cost of hired help has skyrocketed and the oil patch has taken a great many of our farm sons.

Over the years trucks and I have developed a hearty respect for each other - so much so I feel I'm qualified to pass on a few tips to any novice drivers reading this book.

To begin with, it's wise not to volunteer for the job at hand, for then you might have to explain the odd mistake you might make while driving.

Then remember trucks, especially old farm trucks, don't steer like cars. You find you have to pit your strength against the wheel, and when you do get it turning, you have to keep on for at least three rounds before the wheels get the message. Keep all this in mind or you'll end up in some neighbor's prize field of oats.

Gears - now they are trouble makers. First and foremost you must not grind the gears. And that's easier said than done, for the simple reason there's four or five of them, all in different positions and all equally hard to find.

However, if you do grind one or two don't panic - it happens to the best of drivers - just don't let the men hear you. There's something about grinding gears that shakes a man up and at harvest time they are usually shook up enough already.

Once you've mastered the steering and found where the gears are (every truck different), you have to move on to signalling. Now every person who drives a combine has a different set of signals for the trucker trying to pull in beside the combine. It's just about impossible to learn them all, so what I've had to resort to many times was aiming for the third snap on his shirt as he guides me in - don't panic if you bump the combine - I know this machine costs about ten times what a car does but they are quite sturdy. I've connected with ours the odd time and other than giving Bill a heart attack there was little damage.

Another no-no is riding the clutch. This also upsets the experts. The giveaway here is the smell of burnt rubber. I remember being asked if I was riding the clutch and with my, "Heaven, me?" was enough smell to knock you over.

Of course there are other times you'll find yourself behind the wheel of a truck. Like backing into the loading chute. This takes a while to master but don't worry - the truck is usually old and the chute is sturdy and will stand a knock or two.

Also, be ready for anything. I remember one time driving the old silage truck while Bill shoveled it off. Everything was going great 'til I gave a turn and found myself holding the steering wheel in mid-air. It had come right off the steering column, but a farm wife has to be ready for such emergencies so I jammed it back on and kept driving.

So just keep calm, friends, and enjoy the moment. Remember most of the trucks we have to drive are the old wrecks ready to be traded any way.

Harvest Time

Harvest time! The combines gobble up the grain swaths and the trucks rush forth to receive the golden dollars. The women, God bless them, are tossed from pillar to post trying to keep all operations mobile while still maintaining their sanity ...

A most delightful and interesting time of year it is! It's a real ball, taking meals all over hell's half acre. The box of hot food sloshing in the back seat is music to my ears. Even the coffee sloshing over the custard pie can be exciting, I tell myself!

I truly enjoy sticking to my kitchen floor as I walk across it - especially when unexpected city guests arrive in white linen slacks. The very sight of dirty clothes fails to sway me - it's harvest time and everything stops for harvest, I've been told!

Then, of course, there's the delightful job of running here and there for machine parts of every size and description.

True, the orders come at some inopportune times like 7 a.m. or 10 p.m. But keep in mind you get the chance to meet all the machine agents within a radius of 50 miles, and you can even inspect all the boxes of bolts in the whole place while waiting. There must be some hidden pleasure here also, I would imagine...

In town, the elevator agent greets you and your yet another sample with a groan, "Oh no, not you again!"

The straw lodged in your muffler falls off on Main Street along with the muffler! It's little wonder farm queens have fall complexes!

The grocery stores are doing a landslide business. You can fairly see the dollar signs in their eyes as you fill yet another cart.

I've proven that the very best time to get rid of moldy jam, soggy pickles and that apple sauce you happened to salt instead of sugar, is now! Slap it in a dinner box and they'll eat it. It must be the fresh air or something, but there's nary a complaint.

Don't forget the cutlery though! Even the stoutest oat straw fails to hold chocolate pudding, and a jack knife and three bobby pins leaves a mite to be desired when the men are in a hurry...

But after it's all over, your bruises are healed, and the new dress is in the closet, you just might realize it was all worthwhile!

Say "Cheese"!

When I was busy raising kids, one of the biggest pains of all was the annual picture-taking each fall at school.

In my day the teacher brought an old Brownie camera, told us to say cheese and took two snaps. It was all over in three minutes. But no more. Now each kid has a separate one taken, plus each class and each team they belong to at the moment. When they finally bring these pictures home for you to buy there's about twenty of each kid in every size imaginable. Speaking as a mother who has enough pictures to paper a Masonic Lodge, it's enough to drive one insane.

But still comes the word you are to have your kids' pictures taken at school. If you don't you're a rotten parent. If you buy only one snap you are really a rotten parent and if you tell them to take them back and say "No way!" then YOU begin to feel like a rotten parent.

It's such a fuss! I think that bothers me as much as the cost. You're living with the kid. You see him too much as it is. You know he's growing up - why else would his jeans come to his knees five times a year? Sure, he'll soon be gone and the pictures will bring back sweet memories. Sure they will, but I have dents in the furniture and a box of old boots to remind me of him! Besides, now that my kids are all married I see more of them than I did when they lived here.

But from the first of September on, all kids are conscious of the fact that it's due time for school pictures.

"Mom, will this pimple be gone in time?" Try and tell her five thousand pimples will have come and gone by picture-taking time. Still the mirror is watched at each opportunity, and when the face is clear she acts like she's had a reprieve from solitary confinement.

Another bugbear is clothes - "Mom, I need a new skirt!"

"Mom, I need a T-shirt to go with my skirt!"

"I can't wear white - it makes me pale!"

"I can't wear green, it makes me look sick!"

But you can bet your bottom dollar any "perfect outfit" needs wash, or a mend or belongs to her sister.

By the time the great day arrives, everyone is in a tizzy. They all dress to the hilt, spending hours in the bathroom. But guess what? Ten seconds before the school bus is due, they rip upstairs and change - usually into their oldest jeans and the blouse they wore yesterday.

Then there is that dreadful day of the big return. I've had as many as five kids coming home bearing that little brown envelope.

"Hey Mom, can I keep all of mine?"

"No, absolutely no!"

"You don't love me, I must be adopted, you NEVER keep mine!"

"Look, I could buy a film and get 24 snaps of you for five bucks!"

But all in the same breath the kids are alternately loving and hating their particular photos.

"Oh Mom, I look hideous!"

"Mom, can I try again - he's coming back in October!"

I don't know about the photographer, but once a year was all I could stand!

Move Over, Dolly!

One of the many, very important issues confronting girls when I was young, was how many inches we were around the chest. It was much more important than how many brains we had. Our shirts just had to stick out in front - the more they did, the farther up the junior social ladder we'd climb.

Every girl come grade seven would run to her respective Mother with the complaint that she was the only kid without a bra. Now if she were the only girl without a pair of shoes or even a sling shot - but a bra?

When I was of those tender years I was built like a popsicle, and even the smallest of these creations hung low on me. But as long as it was hooked at the back I felt like a grown woman. After all, you just weren't with it if you wore an undershirt. An undershirt was as bad as being caught wearing your brother's underwear!

The odd kid was lucky enough to actually fill out a 32AAA, but most of us had to resort to filling the empty spots with something else other than ourselves. Kleenex was the best, but in war years there was little money for such things as Kleenex so we had to give a good number of healthy sneezes before our moms would allow us a Kleenex. It was easier to use toilet tissue, so we'd stuff away to our hearts content. Of course the toilet tissue was of an inferior quality then also, and it was not unlike stuffing thistles down our fronts. But beauty was far more important than comfort. The only drawback was the rustling! In fact it sounded like a wind whipping the willows when all us girls swept into class each morning. Of course by evening the rustling had quit, but by that time we were sagging.

I'll bet our school division had the highest toilet tissue bill in the whole province!

Our one aim in life in grade seven was to strain the buttons on our blouses, but of course times were hard, so our moms used to buy our clothes big enough to do at least two years. By the time the two years were up they strained okay, but at first we had to permanently strain our backs to stick out at all.

I remember stuffing mine with Dad's hankies. True, they

didn't rustle, but they were the red and white dotted ones and gave my front a strange spotted effect.

As we grew up, other things took priority and nature added a bit.

I've now graduated but now they tell me it's the flat look that's in. Who says every dog has his day?

Don't Marry In The Fall

Anniversaries are beautiful. Everyone knows that. A special day to be celebrated in a special way. Are you kidding? Down on the farm it's usually the very opposite.

If you were as smart as I, you probably got married in harvest season, haying season or calving time!

We got married in October after an early fall, so our anniversary was usually spent combining, chasing cows or some other nuisance. One year I remember well - I had made my mind up that this anniversary, we were going to the city for overnight. I started planning two months ahead - you see, it takes time to plan things out on the farm.

"Bill, do you think we can get off this place this year for our anniversary?"

"Gee hon, I don't see why not, but you know farming!" That to me was a real committment. Men, you see, don't plan ahead too far, especially farmers. They can't because so many things can intervene. But women are different. They plan, and plan and plan. So not being any different I started planning. I dusted out my suitcase, bought some Wind Song cologne and looked through my closet.

We started harvesting early that year and all looked rosy. Each time I took a meal to the field I'd ply Bill with questions, hoping for assurance.

"How's it going, honey?"

"Round and round, Bets."

"Do you think we'll be done by the fifth?"

"Now don't go counting on it, hon - you know harvest weather is unpredictable!"

But I kept right on figuring. Fifty acres a day, five quarters in crop. Boy, I hope he has a spare moment for a haircut. But just as we'd be going nicely, clouds would gather and spit defiantly on my hopes.

Then the sun would venture forth and off I'd go, planning once more.

There would be a break in the combine and I'd be tearing to

town for the needed part, but even then I kept hoping. Maybe by some sheer miracle we'll combine five hundred acres in one afternoon!

I even went to town and bought some Este Lauder Youth Dew. I felt I needed some extra youth at that particular time!

The first of October arrived, the second and the third but the combine was still running behind schedule. But I could still pray for a rain or even a snow - and I did, but God must have felt I was being a bit selfish, for October fifth arrived sunny and warm.

The dinner hour I had set aside for roses and candlelight found me under the combine holding a flashlight and a wrench, yelling at Bill not to hit my thumb! And the romantic evening I had envisioned found me sitting in the grain truck reading 'The Egg and I' by flashlight.

We did get away later in the fall, so this tale doesn't have a CBC ending after all!

Sitting Duck

I love hunting and I love eating wild game. But I'm not so keen on the in-between preparation - especially wild birds.

I will never forget the first time I had to clean and pluck two big Canada Geese.

I was a bride at the time and figured the best way to keep my marriage happy was to attempt any little chore which was set before me, whether I wanted to or not.

A friend popped in one day and nonchalantly tossed two big geese on the cupboard with a "There you go, Bets - goose for supper" and off he went, whistling away.

Quelling the urge to throw them out the door after him, I eyed the creatures.

Well, I had plucked chickens so I felt these should be about the same. I rolled up my sleeves, put on a pot of water to boil and got to work.

Using gallons of water, heating pot after potful, I found they weren't co-operating at all. With each feather I plucked out, seventeen more seemed to take its place. The feathers I did remove stuck to me, the floor and the window above the sink.

Pluck and scald, on I went. My nose was itchy with the fumes and feathers, and the whole of the kitchen looked like an open eiderdown quilt.

By two in the afternoon I still had fuzzy little patches left on the bodies and I looked like I had been dipped in feathers. In final desperation I ripped off the fuzzy bits and laid them out to clean.

I managed the cleaning without using a tablespoon and proceeded to stuff them. The final affair was a sight to behold. Two bedraggled bodies which looked like they'd come through the Battle of Waterloo, with slices of onion and twigs of parsley sticking out here and there.

I popped them in the oven with trepidation and got to work de-feathering the kitchen.

When the men came in for supper I brought forth the fruits of my labor.

I wasn't really expecting any great praise for my culinary efforts but I didn't expect the reaction I got. The hired man just sat there and stared. Bill sat back in his chair and gasped, "What on earth is that???"

"Can't you tell stuffed goose when you see it?" I snorted self-righteously.

Still, I could hardly blame him for his concern. The creatures adorning the platter looked like anything but a well-dressed goose. They looked rather like two beat-up hamburgers with wilted lettuce.

But they tasted great - and that's the main thing, I figure. After that affair I still hunt but I made a rule that very evening - whoever bags the game prepares it!

Heat The Irons, Boys!

Branding time - isn't it exciting?

A lot of my friends figure branding day is a snap. Why, just slap a few brands on the calves, give them a pat and all done. But it just isn't so. The only normal thing you'll do all day is get up. From then on it's sheer bedlam.

You see, any cattle can raise Cain with the most orderly of souls when it comes to branding time. First you have to bring the cattle in from the range. This can be easy but it also can be a real circus. If the men with horses can't get them turned in then all the kids, a dog and the housewife will have to get in the act. I have found myself out chasing the cows out of the bushes, broom swinging, more times than I care to remember. You see cows when out in the pasture, don't get too much excitement so when they see this entourage coming over the hill they decide to scamper about making their own rules as they go. The great escape is enough to send even the most stout-hearted to the house in a snit.

Once the cattle are all in, the wife finds herself tearing to the house to prepare the coffee for the men before they even start the job.

After pulling thistles from her skin and dabbing shaving lotion on her mosquito bites, she calms herself - sort of the lull before the storm. For then she may be called on to help separate the cows from the calves.

Usually this routine means the boss reads the numbers, a neighbor mans the gate and the wife and kids do the chasing.

Of course the branding itself is a joy for the men. At every branding, friends come from miles around, and everyone - be they short, fat, tall or whatever - thrives on brandings. I think when a man has to do a job he absolutely loves to do he's got it made. And so it is with branding. The men who love horsemanship will do the heeling, each one trying to outdo the other. Next, all the others congregate to hold down, brand, vaccinate and act the mighty surgeon. The operating room is the corral and the head nurse is a snuff-chewing old cowboy, his gown a worn G.W.G. shirt. One thing about branding - it brings all the friends and neighbours together, and all the women dig in and help.

The cook keeps an eye on the lane, popping in an extra spud as each car arrives. Actually the yard takes on the appearance of an old-time fair. The kids are ripping about and the old-timers are sitting on the corral remembering by-gone times.

By the time it's all over the food is set out and the liquid refreshment stands ready. The aroma of rich, dark coffee intermingles with the smell of singed hair, and the dinner music is provided by the mommies calling for their babies. Neighborliness and friendship are as obvious as the smell of creolin, and stories are told and re-told.

The gang is reluctant to leave but as they drive off in their old pick-ups, sore and stiff, they think ahead to the next branding, an integral part of farm life.

Common As Dirt

"Common as dirt" - that's how the old saying goes.

Dirt is very common, especially if it's mixed with a liberal dose of rain.

The mud is just everywhere. Back porches are so deep in mud you could plant your sweet peas in it and you daren't scrub up, simply because you're likely to bring on yet another shower!

There isn't a clean boot to be found on the place and you don't dare step out the door in shoes - heaven forbid!

A great many farmers have had to resort to the good old plank system, placing them strategically here and there. But have you ever tried walking one? I really believe it can't be done.

No matter how far apart or how wide they are you just normally tip. I think our anatomy just isn't built for the chore.

Another thing it's impossible to do is carry a new calf along a plank while heading for the house for a warming up. These little creatures are slippery and energetic and many a good man has met his waterloo trying to prove it can be done.

Kids are drawn to puddles like bees to honey, and a mixture of one kid and a puddle is a terrible combination. It's really funny, you know - any good kid can make it to the middle of the puddle but they rarely can make it back.

How many times in the spring have you heard their yells of distress?

If there were only three sentences in our vocabulary, come spring they would be "Good morning, I love you and Help Mom I'm stuck!"

Another thing that's a nuisance is the fact the men and kids go through clothes like candy. One set of jeans a morning at the very least. They get them so stiff and muddy they can stand up by themselves and you'd swear the poor guy was still in them.

The poor old washing machine can barely keep up with the extra loads. Just peek in - you'd think the gyrator was digging an addition to Diefenbaker dam.

Probably the worst part of all is the way it draws visitors.

Now don't get me wrong. I love people but I do wish I'd get more when I've got my back porch clean!

I made a dreadful mistake one year and promised I'd hold the ladies' guild for April. It should have worked. I was through my spring cleaning, the weather was beautiful, but oh, the yard!

It was a sea of oozy mud. We laid out planks here and there and hoped for the best. But good heavens, all too soon the cars started arriving.

I spent the entire first hour pulling, pushing and cajoling women over planks to the safety of the porch. Half of them didn't have boots on because of course it was dry in town.

But then one must remember - it's right that good church ladies should walk the straight and narrow, but we shouldn't expect them to walk the plank!

Homemakers Don't Work!

I learned something new this week. We homemakers don't work! That's right! I had a call the other day and it contained this message, "Oh, you don't work, eh - you're just a housewife!"

Gee, I was almost sure I worked the odd time around here! But no - we don't work, girls - and that's that!

Still I wonder just what I've been doing for the past twenty-odd years of my life. I got awful tired sometimes - and disgruntled for someone who was just fooling around.

Apparently I've been vacationing and didn't have the sense to appreciate it - and imagine me feeling guilty about the precious time I thought I was stealing from my work day to follow my hobbies.

I guess it isn't everyone who is fortunate enough to have the fun, fun, fun of looking after a husband, five kids, an assortment of hired men, cats, dogs and other animals on an old homestead - and I've been under the illusion I was working just a bit!

How lucky I've been scrubbing floors, wiping up spills, cleaning cans of jam off the walls and toothpaste off the bathroom ceiling. And what about all the joy of getting five kids onto the school bus day after day and then wracking my poor stupid brain over that new math, the science fair and untold extra activities all kids get so engrossed in.

And all the washing and ironing I've done five thousand times through the years - even quelling the odd notion I had of dumping the whole mess in the creek and sitting down with a good book. But I really was having the best of everything, a holiday every day, and didn't even realize it.

Of course if I'd known I was merely vacationing I wouldn't have screamed bloody murder when the kids beat each other over the head with the baseball bat, or tossed my hands up in despair when the baby of the year threw her corn flakes and milk all over my clean floor.

Here all the while I've been impatiently waiting for bedtime and counted the quarter-seconds 'til each little Kilgour started school - you know, only five thousand six hundred and ten and one-half! And then when she was finally launched turned proverbial cartwheels on the front lawn - forgetting all about my arthritis!

I did enjoy watching their little antics like popping the old tom cat on the dog's back - so I guess I did vacation a little bit, but all the while I was under that old illusion I was working - fancy that!

I can't wait to tell Bill that I'm sorry I got so snitty the day I ran over his fool scoop shovel with the grain truck and he had the nerve to blast me - here I was just cheap labor. But the truth of the matter was it was a joy to be out there in his dirty old field - as it was when I hauled chop, chased cows and fed men. But how was I to know all the time I was hoeing the spuds that I wasn't what you could say, working! And it took a city girl to enlighten me.

Now I'm not a rabid women's libber. I won't burn my bras - I need all the support I can get! Nor am I a man hater - how the heck do you think I got into all this mess in the first place? But the next career girl who tells me, "Oh you don't work - you're just a housewife!" had better duck, or she's going to get belted in the chops with a jammy dish cloth!

The Queen's Mail

Since time began, mail delivery has been of the utmost importance. When mail was delivered in saddle bags by horse and rider, the motto was "The mail must go through," Rain, hail, sleet or snow couldn't stop delivery of the King's mail.

It was a good thing it always was delivered, because millions of people depended on it. To the early settlers it made life worth living, sometimes when there was little else to make it so. It may have come only once a month, but how they looked forward to it. Parcels from Timothy Eaton's containing material for the new dress for the barn dance, or wool for the kids winter sweaters and mitts. It was often the only outside link with the rest of the world and to many settlers from overseas it was the only way to keep in touch with the folks in the old country.

No, there was no talk of mail strikes in those days!

I remember when I was a kid, mail time each day was a social event. Most everyone in town headed down to the post office before the mail even got there and waited 'til it was all sorted and the wicket went up. Each letter was collected from its box as it was popped in, and read while we waited for more.

On Saturdays all us teenagers would sit around the wide low window ledge while we waited impatiently for a special letter to come our way. If we weren't expecting one we'd go down anyway and just pretend. When one did come we'd dash to the box, twirl the right combination and retrieve it with a great flourish.

While the sorting was being done everyone would visit. The parents would talk about their kids and the kids about their parents. It would be at the post office one would learn of the outbreak of measles and the latest gossip about who was running around with who. And whose kid it was that was stealing Jake's crab apples - of course us kids knew who it was all the time, as he had been selling them at school recess all week!

If a lady had recently had an operation it was at the post office she could explain it all. The old men could discuss the political situation and teenagers too shy to speak to each other could 'just happen' to meet at the post office. They mightn't even speak then but they could gaze at each other - you see, it was a romantic meeting spot as well as a post office.

It's funny - the wages were terrible back then but every kid wanted a part-time job at the post office. It was truly the hub of the community!

Mother's Gone A-Hunting

For me, the most thrilling and satisfying trip to go on is an antelope hunt. Each year, if I can work it, I indulge in this exciting hobby. I choose the antelope hunt especially because you can usually bag one in two or three days, thereby enabling busy mothers to go on a trip not otherwise possible.

The fever hits me about the end of July when I'm busy taking cool lemonade out to the very hot men in the hay fields. It must be the sweet smell of the hay or something that sends my thoughts south to antelope country, for there they go!

When I've acquired the better half's consent to take his dippy wife on yet another hunt, I send for the special applications one must fill out for the proper license.

In Alberta you must send in applications and then they draw for the lucky hunters. I imagine the government feels it's keeping a better control on the amount of antelope taken out by hunters, but I have always felt more hunters apply and possibly more antelope are shot than would be if the season was just opened. Something like prohibition - if you can't get it everyone wants it.

However, once the applications are sent at the end of August, the long wait begins. The suspense is terrible!

Finally about a month later the lovely green license appears and our particular zone is south of Medicine Hat down in the sage brush country.

Now the planning begins in earnest. I stash away little goodies from each week's groceries. On these trips I have vowed the only cooking I'll do is whack off slices of bologna.

Warm clothes, greasy face cream and sunglasses are a must. I hate being cold, sunburnt and blind - in that order! As I plan my apparel we start out practising with our rifles. My particular one is a 30-0-6. This gun has little kick and it will ensure a quick painless kill. By the time October 22 comes around, the old coulees back of the barn look somewhat like a pot cleaner and my shoulder has taken on a hue with every color of the rainbow represented - but boy, can I shoot!!

The week before the trip I start hooking dear friends and kind relatives into keeping the five kids. How they love to go! Just think - a whole four days without a screaming Mom and a sarcastic Dad. Sheer heaven for all!

The great day arrives. Bill has spent all his spare seconds reloading shells, and as I pack the suitcases he picks up all the other gear needed. We always camp out, so a tent with all accessories is essential. I also hide away the gifts for the children for they fail to realize there are few shopping facilities out on a sagebrush plain, and that's where we're off to.

The car loaded, all kids free of illness - touch wood - and the weather is beautiful.

I hold my breath for the first forty miles for fear that something will hinder our getaway, but we make it. How lovely the country is this time of year! The bared fields with just the stubble remaining, the bales piled in readiness for pickup and the wild ducks settling on little sloughs. I soak the tranquility in like a sponge.

When we arrive at our particular zone we find a neat little nook for our camping spot, away off the road near a hidden meadow. While Bill hauls out the tent and sets it up with the precision of a surgeon, I tear about with tent pegs, rods and hammer, tripping over the ropes as I go. After all is up and secured we have a snack - Bill's favorite beans and my meatballs and gravy, fruit and cookies washed down with hot coffee. Food for a queen and I feel like one too, perched on our sleeping bags with nothing to do for four days but act the mighty hunter.

If there's time before dark we look over our terrain and possibly spot our game.

As we enjoy the sunset Bill fills me in on all sorts of information. Grass, shrubs, tracks. He's an authority on all and I'm duly impressed, for seldom do we get this chance to talk without being interrupted!

Back at the tent we plan our strategy for the next morning and line up all rifles, clips and ammunition in readiness, because we must be out at the first crack of dawn.

Bill is snoring peacefully away while I mull over many thoughts. "Will we see game?" "Will I hit what I aim at?" "Will I beat McKibbin this year?" - (McKibbin being a good neighbor and friend as well as a crack shot and a terrible tease.)

Morning comes and as I crawl into my red everything, I dodge the mirror on the tent pole, for well I know what I look like at 4 a.m.!

We start stalking our game. The excitement runs high but quiet. Above all, you must be quiet! At Bill's instruction we drop to our knees for a real sneak. Just maybe there might be some in that meadow over there behind this hill. The sun is just lifting over the horizon, a red shaft of light.

Next move - that rock. "Belly down, Bets," Bill orders, "Keep the barrel up." So I lift the barrel of my 30-0-6.

As I reach the rock I peek over the top. With a gasp I stretch further. The whole valley is alive with antelope!

Bill yanks me down from my mesmerized state and looks over for himself. But they also spot us and spook! As they run off to the sheltering ravine Bill orders, "Take a big buck, Bets."

While I try and get a bead on a likely buck Bill is having five fits beside me. Being from the old school he feels honor bound to let this dude hunter have first shot. Finally as I pull the trigger on my buck he lets loose on his. As I see mine go down I start jumping.

"I got him, I got him, at 200 yards I got him!!!" Jumping up and down like a mad woman.

"For gosh sakes, quit yelling! Someone will think I shot you!!" Bill declares. "Don't forget your empties." (To be filled for yet another hunt.)

We wound our way down to our game. Bill had a big trophy and had pulled off a wonderful shot. Mine was a big animal - not a trophy by any means, but to me it was magnificent! Enough so to start me yelling again, and my long-suffering Bill was just as delighted as I, a huge grin on his Scottish face. We both had our game. We duly tagged them and Bill went off for the car. I stayed to guard my prize but I can tell you ten elephants couldn't have pried that buck away from me!

After taking pictures, which took longer than the entire hunt, we headed back to camp, our trophies in the back.

After a belated breakfast over gun talk and smug chatter, we took our antelope to a friend's garage to hang while Bill and I went on a bird hunt and a sidetrip to Maple Creek, just over the Saskatchewan border. There we found beautiful old buildings which had been hauled in piece by piece from the States before the turn of the century. We generally just made good use of the remaining days.

Our hunting wasn't quite over, however. The last evening, upon returning to camp we found a mouse chewing away on a crust between the two sleeping bags.

Well, my Bill may be a mighty hunter, a fearless farmer and in general a very brave lad, but at this particular time he let out a mighty yell, leaped back on the sleeping bag and standing there on one foot hollered, "Get it Bets, get it!!"

It was a scene from a movie - Bill shouting, me chasing around with the fry pan hitting where the mouse had been and the poor mouse running here, there and everywhere. Finally it managed to escape out the tent flap, where it probably expired from fear or an ear rupture.

I got Bill down and aspirined, and peace returned to our wee camp.

The next morning we packed up and home we went. The end of a glorious holiday, a dandy hunt and a real week-end.

As the hunt ended the stories began - enough to last until next year.

Housekeeping The Easy Way

Are you a perfect housekeeper? Well, if you are, turn the page. Are you a hit-and-miss housekeeper who finds it difficult to keep the place straight and get everything else done at the same time? Good, read on.

First and foremost, get over your guilt complex. You're not along in your dilemma, and just think of the boost you're giving all your friends when they catch you in the muddle. There's nothing that cheers a farm woman more than finding another who's a worse housekeeper than herself.

Now throw out all dust covers, plastic on chairs, covers on everything from the mix-master to the toaster. Who ever heard of dusty toast? And it will be that much less you'll have to shake, wash or iron. This goes for mats too - all except the kitchen one that catches the drips when the men wash up. It also comes in handy for shoveling the dust under when you've burnt your dust pan yet again.

Drawers are my delight. In drawers you can stuff a wealth of mess and it helps keep counters clear - if your counters are clear, people will figure you're a great little manager.

A warning. Don't leave your husband's treasures - his tomahawk, forestry maps, or his old navy shells - on top of a cupboard. He will always pull them down to show them off, giving a hearty blow to dislodge the inevitable dust right in his guest's face and to your great shame. Retire all such items to drawers as well. You can use that spot for the hornet's nest the kids take to the Science Fair. If cobwebs are wrapped about it, all the better for the scientist in the family to explain in class.

Now for those thought-provoking finger marks and jam smudges which every house is blessed with when kids are little. Wipe the most obvious the minute you notice them and in a week you'll have them gone and probably replaced with four more, but at least they're new ones.

Burnt pots and pans can be a pain. They are ugly and noticeable in the sink, yet have to be soaked - so just pop them on the top step on the way to the basement. Watch for the day the gas man comes, though. It would be a mite embarrassing having to dislodge a size-ten work boot from your pot.

Ironing is another nuisance. You can't hide it in the bathroom or the kitchen, but I have a friend who hid hers in the back seat of the car when unexpected company arrived. Her husband spoiled everything though by taking the male visitor on a tour of the crops - in the car! (What I have done in an emergency is hide it in the deep freeze.)

When planning on entertaining, don't ever scrub for five days ahead and then be so tired you can't enjoy your guests. Draw the drapes, shove stuff under the couch and light candles. The effect is soothing and romantic and hides a multitude of sins, so to speak. You'll have to clean the house the next day regardless, and this eliminates one set of cleaning.

Also, if you serve tea, dust out the tea cups. This you must do or you might have to watch some poor guest trying to dislodge a ball of lint from her front tooth, and this doesn't do a lot for friendly relations.

Hide all daily muddles behind the couch if you get a phone call informing you of a quick surprise visit - shoes, gum wrappers, jars with caterpillars. All mothers will know what I mean.

I had all our kids trained - while I showed the guest the flower bed, they tore around like crazy stuffing junk out of sight.

The garbage can is still a housewife's best friend. You can solve many a problem by tossing it in the garbage. The cake that flopped, the 1920's curtains your mother-in-law gave you, the pair of jeans you spilled the bleach on.

And if things get too deplorable at home, try to visit a housekeeper who has even more of a mess then you. This will keep you going for a few days 'til you finally get a moment to swish out your house. When you do feel the urge to clean, do a complete job - every nook and cranny. Then while it's shiny clean invite all the folks you want to, especially the fussy ones. This may inspire you to keep it in such shape for a few days. Then again, if you're like me, emergencies will arise and the man of the house will bring in a sick calf to be wiped and warmed, and off you go again - that's when these tips I've left with you will come in handy!

Skunk Trouble

Oh, am I in trouble - skunk trouble. One little striped creature has taken refuge at Crocus Coulee and finds it so absolutely exciting he refuses to leave. The whole farm is at his disposal. The dog's dish, the garden, the bale stack and under the back porch are all his domain.

He's about half grown and is kinda cute, but the smell! The poor dog has given up. He's tangled with him three times and after three sprayings and having to endure the bath and the absence of all love and affection from the human element for days, he's staying clear of the visitor. I even saw the skunk eat from the dog's dish, and he just slunk away from it.

I'm a little more stubborn than Duke, though. I haven't been sprayed yet so I'm a little braver. My main method is to try and shoot the skunk. I've shot a good box of 22 shells at it and haven't even scared it. I've even tried Bill's 243 but I forgot to allow for the rise or fall or something, so I even missed with it.

Now I've shot the odd antelope and tons of gophers, so I must just have what old hunters call, "buck fever".

I think after all this missing, he's really enjoying the whole war. Each and every day he saunters across the yard and each time I take aim. However, I've got to change my tactics somewhat - my last shot nearly set the hired man back ten years. It went whistling by so close I nearly burnt his shoe laces when he inadvertently stepped out of the bunk house at the wrong time.

The final straw came last night, though. I smelled the creature under the verandah and even down in the basement. Now it's one thing to see it across the yard when you're in the porch with a gun, but to smell it in the house when it's dark outside is another.

Each time I went by the basement door I tiptoed, and I didn't dare open the front door. It really got to me and I did what any red-blooded wife would do - I took three aspirins and yelled at my husband.

"For heaven's sake do something about that skunk!" The reply was,

"Bets, do you know what you do with skunks at ten-thirty at

night? You let them sleep!" and he did the same. I'm not through yet, however. I hunted up Ian's old trap and I'm reading up on what you bait a skunk with. That should do the trick. Either that or you'll meet me walking down the lane, suitcase in hand - I just hope I'm not taking the aroma with me!

The Science Fair

If I have one nightmare left over from raising five kids, it's dreaming that they're all home once more and the next day is the Science Fair! What a headache that was each year.

I realize these Fairs are a wonderful project, teaching many things to many students. I also realize it gives the scientifically-minded children and their clever parents an extra opportunity to fiddle with formulas, fix mighty motors, and build some great objects. I hold great admiration for these people.

But what of the student who's about as scientific as a wet noodle and has inherited this lack from equally unscientific parents?

For days, weeks and months this dilemma whacked away at our otherwise happy home. What to make was the big issue. The first dream project was always grandiose: a life-size oil well, a new chemical to kill wild oats or a space station over Three Hills. As each of these ideas went down to defeat for obvious reasons, others took their place. A battery-operated potato peeler or a mechanical clothes picker-upper.

As each idea gives way, so does self-confidence. But each idea, before it could be eliminated, required supplies to make it with, so there would be ten thousand trips to town for materials and ten thousand more to and from the kid he was working with.

Materials caused trouble. You find yourself buying Cheezies for the yellow bag, brussel sprouts for the green basket and anything from bath oil to liquid soap for the container.

Each and every effort is thwarted and then it's up to Mom to come up with a brilliant idea. But then Mom is not that brilliant either. Now if it was a chocolate cake that was needed or how to mend blue jeans I could probably help. Or how to plant a garden or shell peas - but no, it has to be scientific.

Usually I sought through the house for something unusual. A Massai spear, Bill's navy hat or the old hornet's nest I kept on top of the cupboard. But the fuss this all caused! Multiply all this by five and you have a general idea of what went on in this house. Five unscientific kids shouting and crying through the weeks at poor frustrated parents who felt like wringing the science teacher's neck - meanwhile, the science teacher was wishing fervently for a good two-week case of the flu.

Now I can crochet and knit, cook and sew and even hammer a nail in if I have to, but I never got higher than a C in Science and haven't improved with the years. The only time I'm not going to allow my grandchildren to stay overnight is near Science Fair time. I'll let their parents enjoy that event themselves - I wouldn't deprive them of that particular pleasure for a moment!

Chinese Sign Of War

I learned today that the Chinese sign for war is a little house with two female figures under it.

It really stands to reason. Women are set in their ways and to have two of them both running a house would mean nothing but war. I believe if you put two absolute angels in a kitchen and told them to run the house and get along, you'd have fire and brimstone pouring out the windows in a week.

A woman's home is her castle and she has complete charge of it. Just show me two women who do everything the same. There isn't such a couple and if there were they'd have something else to disagree about.

I know for certain I wouldn't want anyone else running my home with me. And why should I? If I want to be grouchy in the morning, burn the toast or boil the porridge too long, I wouldn't want another woman telling me I shouldn't.

If I have the habit of ripping the cereal box in the wrong spot or squeezing the toothpaste in the middle I wouldn't want someone telling me I was wrong - I already know that - and I wouldn't want anyone else to know.

If I toss the eggs in whole when a recipe tells me to well-beat them, that's my business. I'll even not sift the flour if I want to be lazy.

If I use my good butcher knife to pry out nails that's my business too. If I use a rolling pin to pound them back in, that's also my affair.

If I want to sweep the last bit of dust down the cellar steps, that I don't want another woman to see.

Now the bathroom is also my domain and mine to keep in any way I want, harebrained or not. If I plug the toilet when I flush the dead canary down it, then I'll have the honor of un-plugging it. If I wish to keep three apple cores and my typewriter ribbon in my lingerie drawer I don't want another to know, and if I kick the coffee cup from last night's midnight lunch under the bed until I vacuum, well, it's my room.

If people write their names in my dust, then fine - it's my dust.

I'll dust when I want and only then.

I think the reason I feel so strongly about another woman in my house is because I wouldn't want anyone, especially another woman, to see how slovenly I can be. What I do know for sure is that if another tried to run my home with me in it, the domestic scene which would arise would make the Battle of the Bulge look like a Sunday School picnic!

New Wheels

Guess what? I got my wish. A shiny new car has replaced the old beast we had. Isn't that wonderful? You'd appreciate it too if you had worked as hard as I did to get it! A good three years of nagging went into this project and that is hard on the system!

I can't remember when I started wishing for a new one. Possibly the day the old car pulled that nasty stalling trick on me. I was in the process of taking a hot dinner out to the busy harvesters who happened to be working some six miles from home. Just as I turned and slowed up for the first corner it sputtered to a stop and refused to start again.

With the pot roast cooling in the back and me boiling in the front I proceeded to try every trick known to the novice motorist to get the car going again - including flooding it. After both the car and I had cooled down a bit and five hundred cars had swept by in disdain I managed to get it going. This type of action didn't exactly help relations between woman and car, let me tell you!

Then of course there was that fateful Sunday morning when my old friend decided to take what pride I had and toss it to the winds. Bill being away, I had to go to church by myself. I dressed quite properly, got there on time and learned a few things. I went to the door, spoke to the elders, shook hands with our minister and walked sedately to the car - the perfect Presbyterian. As I went to start the thing I managed to drop the key - right through the hole in the floor board. Now any one of you who has ever tried this trick knows full well that the only way to retrieve the key is to get belly down and reach right on the street. Needless to say this act of disgrace caused quite a stir and shook up my equilibrium as well as the astonished onlookers still standing on the church steps.

The next day the minister phoned three times inquiring as to the state of my health. I'm sure he figured I had been drinking anti-freeze under there. Why else would a woman be sprawled under a car at high noon?

And to condone my deliquent thoughts about a new car let me fill you in on my HOP, SKIP, and JUMP episode. I had stopped for a coffee at a friend's house in town. As I walked up the sidewalk to greet her she let out a merry shout. I turned about only in time to see my trusty pal rolling merrily along gathering momentum as it went. I took a dive for it and grabbed the door handle, only to have it come off in my hand. Not bothering to faint I reached for the rear door, got it open, all the while hopping sideways. In I jumped, reached over just in time to steer the car off the dentist's lawn.

This whole episode took only a couple of minutes but it aged me forty years. These experiences mixed in with stuck doors which only happened when I was driving, splutters and splatters up and down Main Street didn't endear me to the old car one bit.

Every time I mentioned the need of a new mode of travel Bill would retort, "Nothing the matter with that Plymouth, hon, just goose it at the corners!"

Well sure, but while I was concentrating on goosing it I forgot to steer and ended up in a caragana hedge.

I'm not sure whether I nagged Bill into a new car or he figured I'd earned one but it's lovely to know I can travel to town and back without little embarrassments - even if it's a bit boring!